DATE DUE

OCT 2 2 1974			
GAYLORD			PRINTED IN U.S.A.

Thermal Properties of Matter, Volume I

KINETIC THEORY OF GASES

PHYSICAL CHEMISTRY MONOGRAPH SERIES

Edited by Walter Kauzmann, Princeton University

QUANTUM MECHANICS IN CHEMISTRY
MELVIN W. HANNA, *University of Colorado*

THERMAL PROPERTIES OF MATTER
WALTER KAUZMANN, *Princeton University*
   VOLUME I: KINETIC THEORY OF GASES

Thermal Properties of Matter
Volume I

KINETIC THEORY OF GASES

Walter Kauzmann
Princeton University

W. A. BENJAMIN, INC.
New York   Amsterdam   **1966**

KINETIC THEORY OF GASES

Copyright © 1966 by W. A. Benjamin, Inc.
All rights reserved

Library of Congress Catalog Card Number 66-13995

MANUFACTURED IN THE UNITED STATES OF AMERICA

The manuscript was put into production on July 23, 1965; this volume was published on February 15, 1966

W. A. BENJAMIN, INC.
New York, New York 10016

Editor's Foreword

ANYONE responsible for teaching a one-year introductory course in physical chemistry must make difficult decisions in the selection of topics he will present to his students. In the past thirty years developments in molecular physics and quantum mechanics have made these areas essential to the student's background in physical chemistry. Yet nearly all of the topics in physical chemistry as taught thirty years ago continue to be of importance not only to physical chemists but also to those in other fields of chemistry, not to mention biologists, geologists, metallurgists, engineers, and medical scientists. Since the time available in the course is limited, important subjects must either be discussed very briefly or passed over entirely, the teacher hoping that the student will be able to pick up omitted material for himself should he need it later in his career.

The same difficulty is faced by the writer of a textbook in physical chemistry. Sometimes it is dealt with by increasing the size of the book, but more often authors compress or eliminate topics they feel to be unimportant. Furthermore, it is becoming more and more difficult for one author to write with authority about all areas of physical chemistry.

The situation cannot be dealt with by arbitrarily choosing a limited number of topics in the hope that they will be recognized universally as suitable for a one-year course in modern physical chemistry. Physical chemistry textbooks should, rather, be designed and written in such a way that flexibility is possible in choosing the topics to be taught in any given year. Different teachers will have valid reasons for emphasizing different aspects of physical chemistry, and any one teacher may want to be able to change his emphasis from one year to the next.

The present text-monograph series is an attempt to deal with this problem. The basic series will consist of about nine volumes, of 150 to 200 pages apiece, each dealing at an introductory level with important topics in physical chemistry. The texts will be written in such a way that the student will be able to learn for himself about those topics that may not have been presented in the particular physical chemistry course he may have taken. It is also intended that in each volume some more advanced material can be included which will stimulate the interest of the students and give some indication of the present status of physical chemistry, not only as a branch of chemistry but also as a foundation for other areas of science.

The total size of the basic series will be greater than that of the average introductory textbook (the basic series will have about 1,500 pages as compared with typical texts having 700 to 900 pages and covering the same topics), but it is expected that, since not all of the texts will be used in any given course, the financial and intellectual burden on the student will not be excessive. Indeed, the intellectual burden on the student should be less than that imposed by any of the introductory texts now in wide use; because of the greater number of pages available in the series, it should be possible to explain difficult points in more detail and yet to reach at least as high an intellectual level as that aimed for by conventional texts.

The basic series will be supplemented by volumes that consider special topics of current physical chemical interest and are written at a level suitable for students in the first year of their exposure to physical chemistry.

WALTER KAUZMANN

Princeton, New Jersey
January 1965

Preface

WHEN heat is introduced into matter, atomic and molecular motions are produced which influence such macroscopic and essentially non-thermal properties as the density, the color, and the dielectric constant. These motions also give rise to entirely new properties (for instance, the pressure of a gas, the specific heat, the viscosity, and the entropy). Chemists have always had an interest in these thermal properties of matter because so many of the phenomena of chemistry are greatly affected by heat. Indeed, the concern of chemists with heat goes back to ancient times, far antedating the birth of modern chemistry in the eighteenth century. Well-trained chemists at all stages in the history of science have always been expected to be conversant with the currently accepted theories of heat. In our day, the study of the thermal properties of matter is an important part of the training of chemists, occupying a considerable portion of the conventional introductory course in physical chemistry. In any of the revisions of the undergraduate chemical curriculum that are likely to come in the near future, a prominent place will surely continue to be given to this subject. The present three-volume series, *Thermal Properties of Matter* I, II, and III, is intended to fill the need for a text that can be used in this portion of the undergraduate training of chemists.

The theory of heat can be approached from two entirely different directions—the macroscopic (thermodynamic) and the microscopic (molecular or statistical). Although both of these points of view are important to chemists, physical chemistry texts have until recently tended to present them relatively independently. Furthermore, the thermodynamic approach has been emphasized and it has usually preceded the molecular approach. Unfortunately, the abstractness and generality of thermodynamics make it rather difficult for the student. On the other hand, experience has shown that the relatively simple mechanical and statistical concepts employed in the molecular theory of heat are singularly meaningful and vivid to the average student. Consequently much can be said for introducing the molecular theory at an early stage in the study of heat and for interweaving the discussion of thermodynamic concepts with the description of their molecular interpretation. Furthermore, the molecular theory greatly extends the application of the thermodynamic theory to real problems and the student can be readily made to appreciate this. This is why, in these volumes, I have made a special point of developing the thermodynamic and molecular theories of heat side by side, so that the thermodynamic theory is constantly illuminated and enhanced by the molecular theory.

The first volume is largely concerned with the molecular basis of some of the important thermodynamic properties of gases—in particular the pressure, temperature, and thermal energy. The molecular theory that arises from this discussion is then extended to the study of those nonthermodynamic properties of gases that depend on the collisions of gas molecules with each other. The second volume will use and extend the molecular concepts developed in the first volume in an exposition of the laws of thermodynamics, most of the applications of these laws being made to gases. The applications of the laws of thermodynamics and their molecular counterparts to liquids, solids, and solutions will be given in the third volume.

It is intended that the three volumes will serve as the basis for a one-semester or two-trimester course which might be a portion of the usual introductory course in physical chemistry. The student is expected to have had courses in calculus, elementary physics, and general chemistry. Although a good many more topics have been included than can be covered in detail by the average student in a typical physical chemistry course, an important aim of these volumes is to provide the better student with additional material that he can study independently. I hope, furthermore, that the average student will appreciate the rather extended and often quite elementary presentations of the abstract basic concepts that must be mastered by all who wish to gain even a minimal understanding of these aspects of modern physical chemistry.

I have always felt that, even at a relatively elementary level, physical chemistry in general and the theory of heat in particular can and should be presented to the student in a logical sequence of arguments, each step following either from previous steps or from clearly understood premises. I hope that students and other readers will agree that I have followed the spirit of this philosophy in these volumes, but on a few occasions it has seemed pedagogically desirable to make use of results which will be justified in detail only later in the study. This is the case with the introduction of the Boltzmann factor and the statistical mechanical expression for the second virial coefficient in Volume I, the proofs of these expressions being deferred to Volume II. I have found that the occasional use of this admittedly illogical procedure does not cause serious difficulties for the average student. Indeed, this procedure has the definite advantage of making the derivations of these expressions much more meaningful when they are eventually considered at a pedagogically more appropriate stage in the development of the theory. The well-known quantum mechanical rules for translational, rotational, and vibrational energies have also been introduced as *ad hoc* assumptions in Volume I. Those students who have been exposed to a course in quantum theory (as, for instance, that given in M. W. Hanna, *Quantum Mechanics in Chemistry*, W. A. Benjamin, Inc., New York, 1965, another volume in the present series of introductory physical chemistry texts) will have seen these rules proven. Once again, however, those students who have not had this exposure will appreciate the proofs all the more when they eventually encounter them—a fact which seems to me to offer definite pedagogical advantages.

An important feature of these volumes is the frequent introduction of exercises—often partially worked out—which are intended to help the student think about what he is learning, teach him how to use it, and present him with some important or interesting special applications. Problems at the end of each chapter should serve somewhat the same purpose, but they are based more broadly on the chapter as a whole.

It is a pleasure to have this opportunity to thank the many

people who have helped me in the preparation of this work. I have, of course, learned a great deal from my efforts at presenting this material to my students during the past twelve years and from many discussions with my colleagues in the Princeton Chemistry Department and elsewhere. I should like particularly to thank Drs. Hugh McKenzie, Victor Bloomfield, Melvin Hanna, and John Ross, who have carefully read the manuscript and have made many valuable suggestions. I should also like to express special appreciation to my wife, Elizabeth, for the care and effort she took in transforming my tapes and almost illegible handwriting into neat and nearly perfect typescript.

WALTER KAUZMANN

Princeton, New Jersey
November 1965

Contents

Editor's Foreword, v
Preface, vii

Introduction   1

Chapter 1   Equations of state of gases; empirical gases   3

1-1   The "primary properties," mass, volume, pressure, and temperature, 3
1-2   The concept of an equation of state of a gas, 8
1-3   The ideal gas laws, 10
1-4   The equations of state of real gases, 22
1-5   Extensive and intensive properties, 44
Problems, 45
Supplementary references, 48

Chapter 2   The molecular explanation of the equations of state   49

2-1   Bernoulli's theory, 50
2-2   The molecular explanation of deviations from the ideal gas law; van der Waals' theory, 64
2-3   Statistical mechanical theory of the second virial coefficient, 75
Problems, 89
Supplementary references, 90

Chapter 3 The molecular theory of the thermal energy and heat capacity of a gas — 91

3-1 The translational energy of a gas; heat capacity of a monatomic gas, 91
3-2 The classical mechanical theory of the heat capacities of diatomic and polyatomic molecules; principle of the equipartition of thermal energy, 94
3-3 Summary of quantum rules for the energies of molecules, 101
3-4 The average populations of the molecular quantum states in a gas, 105
3-5 The thermal energy of a system of molecules, 107
3-6 The thermal vibrational energy of a diatomic gas molecule, 109
3-7 The thermal rotational energy of a linear gas molecule, 114
3-8 The thermal translational energy of a gas molecule, 119
3-9 The complete partition function and the total thermal energy of a diatomic gas, 121
3-10 Simple approximate expressions for the temperature variation of the heat capacity, 125
3-11 Effects of intermolecular forces on the thermal energy, 127
Problems, 127
Supplementary references, 130
Appendix 3-1 Physical basis of the Boltzmann factor, 131

Chapter 4 The distribution of molecular velocities in a gas — 137

4-1 The concept of a distribution function, 137
4-2 Velocity quantization for particles in boxes, 142
4-3 Derivation of the Maxwell-Boltzmann distribution function, 147
4-4 Some properties of the Maxwell-Boltzmann distribution, 154
Problems, 161
Supplementary references, 163

Chapter 5 Molecular collisions and the transport properties of gases — 165

5-1 Collision frequency in a gas, 166
5-2 The mean free path, 180
5-3 The transport properties of gases, 184
5-4 Approximate molecular theory of transport properties of hard sphere gases, 197
5-5 Experimental tests of the hard sphere theories of transport properties, 208

5-6 Refinement of the molecular theory of transport properties, 216
Problems, 235
Supplementary references, 237
Appendix 5-1 Evaluation of the integrals in equation (5-22), 239

Index of Symbols, 243
Index of Subjects, 245

INTRODUCTION

ONE OF the important and interesting aims of physical chemistry is to explain the properties of matter in terms of the motions and spatial arrangements of atoms and molecules. This aim has been more nearly achieved in the physical chemical study of gases at low pressures than in the study of matter in any other condition. (By "low pressures" we mean pressures below, say, a few atmospheres at ordinary temperatures.) The structure of gases at these pressures is particularly simple: such gases are collections of molecules which move randomly in space and which collide with each other relatively infrequently—that is, the molecules are so far apart that much of the time they exert little influence on each other.

In order to proceed from this picture to an explanation of the observed properties of gases certain concepts and procedures are required which are not at all difficult to understand. Many of them (for instance, the concepts of intermolecular forces and distribution functions) are useful in explaining the behavior of other states of matter. Furthermore, the properties of the gaseous state play a role in many important practical processes, such as the operation of the internal combustion engine, the function of the lungs, the motions of the winds across the earth and the flight of airplanes. Gases, therefore, provide a useful and pedagogically attractive starting point for the introduction of students to physical chemistry. The study of gases also gives the student an example of the kinds of problems physical chemists try to solve and how they go about solving them.

This volume describes some of the important properties of gases and shows how the observed relationships between these properties may be explained by means of relatively simple molecular models of gas structure. In the first chapter we shall describe the observed relationships between the pressure, volume, temperature, and amount of gas present in a system. Chapter 2 presents the molecular explanation of these observations. This explanation is based on the notion that a gas is a collection of particles moving at random with different velocities in all directions. At first we shall assume that the molecules are merely points in space which have the property of mass, but which have no influence on each other. Then the model will be modified by allowing the molecules to interact with each other—attracting each other when they are far apart and repelling when they are brought close together. The thermal energy of such a collection of molecules will be discussed in Chapter 3, both from an experimental point of view and from the point of view of the molecular theory. Chapter 4 will show how, using the concept of a distribution function, one can extract a surprising element of order, or predictability, from the chaos of molecular motion that underlies the basic model of a gas. Chapter 5 considers some additional properties of gases which depend in an essential way on collisions between molecules. These properties include the rate of diffusion of one gas through another and the thermal conductivity and viscosity. They are important because they are capable of giving a considerable insight into the detailed dynamics of molecular collisions. Since most chemical reactions are the result of violent molecular collisions, this study has important implications for the understanding of the molecular basis of chemical change—perhaps the most intriguing topic in all of chemistry.

Chapter 1

EQUATIONS OF STATE OF GASES; EMPIRICAL RESULTS

1-1 The "primary properties," mass, volume, pressure, and temperature

Gases are substances which completely fill any container into which they are introduced. Let us suppose that standing before the student on his desk as he reads this chapter is a flask containing some gaseous substance such as air. The air in this flask has a great many properties—color, odor, density, velocity, viscosity, thermal conductivity, electrical conductivity, magnetic susceptibility, dielectric constant, chemical composition, and so on. In this chapter, however, we shall be concerned with four particularly important properties of a gaseous substance: mass, volume, pressure, and temperature. Because they are so important and are so frequently measured by chemists we shall call them the "primary properties." Before considering their relationship to each other, let us be sure that we understand what we mean by these four properties.

The properties in question are expressible as numbers, and can be best defined if we describe how one goes about determining these numbers. (This way of looking at scientific concepts is almost always the safest way of being sure that one knows exactly what a concept means, and deciding the circumstances under which the concept loses or changes its significance.)

The *mass* of a gas in a vessel can be determined by weighing the vessel when it contains the gas, and then weighing the vessel again after the gas has been pumped out. This procedure is straightforward, and although it is possible to imagine some subtle complications regarding the concept of mass, these complications play no role in physical chemistry, at least not under ordinary circumstances.

The *volume* of a gas can be most easily determined by weighing the water required to fill the vessel containing the gas, since the volume of a unit mass of water between 0°C and 100°C is very accurately known. The procedure is simple in principle, but a complication arises if the gas and the vessel are at an elevated temperature, where the mass of water per unit of volume may not be so well known. This complication can be overcome, of course, by using other liquids whose densities are known at the desired temperature. Furthermore, if the vessel has a simple geometry (a cube, sphere, cylinder, etc.), we may measure its dimensions and the volume may then be determined by means of a suitable geometric formula.

A more awkward complication arises in determining the mass of a given volume or the volume of a given mass of a gas when the gas is not in a container—say when it is flowing out of a nozzle or around the wing of an airplane. We shall avoid complications of this kind for the present, however, by considering only gases in closed vessels. On the other hand, we shall be very much concerned with vessels whose volumes can be altered—particularly with vessels in which one wall can be moved—for instance, a gas contained in a piston chamber or a gas under compression by a liquid in a gas burette (Fig. 1-1). Clearly, each position of the piston or of the liquid corresponds to a volume which can be determined in principle in the manner described above.

The concept of *pressure*, although apparently simple in principle, is in practice subject to a good many qualifications. It may be measured by means of a manometer, with which one observes the height of a column of an inert liquid, such as mercury, that is

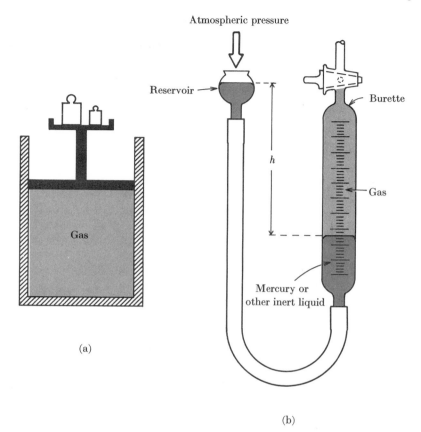

FIG. 1-1 *Devices for observing the behavior of the volume of a gas when the pressure on the gas is changed. (a) Conceptual piston-chamber device. The pressure may be changed by changing the weights on the platform. (b) Gas burette. Pressure may be changed by moving the reservoir up and down and the volume may be read off from the graduations on the burette.*

balanced by the force exerted by the gas on the liquid surface. (This measurement will be described in more detail below, in Section 1-3.[1]). If a gas and its container are at rest, and if they are

---

[1] There are also other devices for measuring pressure. For instance, certain crystals, such as quartz and Rochelle salt, have the property of producing a voltage difference across particular crystal faces when they are subjected to a uniform pressure (the *piezoelectric effect*). This voltage

located in a region free of gravitational fields, then the pressure that the gas exerts on the container walls is the same at all points on all of the walls. Under these conditions one can speak of the pressure on the walls as "the pressure of the gas," and this pressure can be regarded as a property of the entire gas sample. On the other hand, if the system is located in a gravitational field, the pressure on the walls will not be uniform. A gas at the earth's surface exerts a force on the bottom of its container which is greater by the weight of the gas than the force on the top of the container. Strictly speaking, such a gas sample as a whole does not have "a pressure." If the box is small, however, and if the gas is not highly compressed, these pressure differences will be negligible so that one is in the habit of speaking of "the pressure" of the gas despite the fact that it cannot be precisely defined under these circumstances.

EXERCISE One liter of air at room temperature and at one atmosphere pressure ($=$ about 1 kg/cm^2 at the earth's surface) has a mass of about 1 g. What is the percentage difference in pressure between the top and bottom of a column which is 1 m in height, filled with air at one atmosphere and located at the earth's surface?

A more awkward complication arises if a gas moves relative to the walls of its container—as will be the case if the gas is swirling about or if one of the walls (say the piston in Fig. 1-1) is in motion. The pressure may then be quite different on different walls, and again it is not possible to speak of "the pressure" of the whole gas sample. This complication will be discussed in a later volume of

---

difference is proportional to the applied pressure and can be calibrated by exposing the crystal to known pressures. Piezoelectric crystals provide a convenient means of measuring pressures electrically. Another electrical measurement—particularly useful at very high pressures—depends on the fact that the electrical resistance of a conductor changes when a pressure is applied. If a constant current is passed through a conductor, the voltage drop across the conductor varies with the pressure to which the conductor is exposed. A widely used mechanical device for measuring pressure is the Bourdon gauge. This consists of a long, curved tube with a flattened cross section. If the tube is filled with a gas, the curvature of the tube varies when the gas pressure is varied. Thus, if one end of the tube is fixed, the other end moves as the curvature changes, and the displacement of the free end of the tube can be used to measure the pressure.

EQUATIONS OF STATE OF GASES 7

this series. It will also arise in Chapter 5 when we consider the flow of a gas through a tube.

It is common practice to talk of "the pressure of a gas" at points in the gas not adjacent to a wall, where the gas exerts forces only on itself. There is no difficulty in defining such a "local pressure" if the gas and its container are at rest. In a stationary gas the local pressure at any point (i.e., the pressure that the gas would exert on a stationary surface introduced at that point) is the same in all directions (i.e., is independent of the orientation of the surface), and it is usually a simple matter either to measure this pressure by inserting a suitable pressure gauge or to calculate it from a knowledge of the forces acting on the gas. On the other hand, if the gas is in motion, care must be taken in formulating the concept of pressure as a local property of the gas. If a large body of gas moves relative to its surroundings, we can imagine that a pressure gauge calibrated in a static system can be inserted into the gas. If the gauge moves at the same velocity as the gas in its immediate vicinity, it will provide us with a "local pressure" which is physically meaningful. A meteorologist may be able to make such a local pressure measurement in a turbulent storm cloud by sending a pressure gauge into the cloud by means of a balloon. It is, however, hardly practical to use this means of defining a local pressure in a stream of gas flowing at high velocity through a nozzle in a turbine or rocket or past the wing of a jet airplane. Clearly, the concept of pressure under such circumstances may require considerable abstraction and discussion before it can be defined or measured. It is fortunate that we shall not have to deal with these difficulties here.

The *temperature* of a gas is an even more subtle property. We must first assume that two objects can be said to have the same temperature if no heat flows between them when they are brought into intimate contact. For our present purposes we may say that the temperature of a gas is the temperature read on some thermometer that has remained in contact with the gas until the flow of heat between the gas and the thermometer has stopped. A convenient thermometer is a glass bulb to which is attached a uniform capillary, the bulb and a portion of the capillary being filled with some liquid—commonly mercury. The scale on the thermometer may be established by inserting the thermometer first in an ice-water mixture and marking the position of the mercury in the capillary as 0°. The thermometer is then inserted in water boiling at 1 atm pressure and the position of the mercury in the capillary is marked as 100°. The thermometer capillary between the two marks being uniform, 100 equal divisions are marked between them. The scale below 0° and above 100° will be temporarily assumed to be marked off on the capillary with scale divisions of the same magnitude. Obviously this thermometer cannot be used below the freezing point of mercury ($-39°$) or much above its boiling point (357°). Furthermore, there is no reason to believe that a mercury temperature scale defined in this way will be identical with one using another fluid or with thermometers whose bulbs and capillaries are built of different kinds of glass, or of materials other than glass. Later in this chapter we shall find that more satisfactory temperature scales can be constructed in other ways, but for the time being it will be sufficient to make use of the above method of attaching a numerical value to the property of temperature.

1-2 *The concept of an equation of state of a gas*

Consider the following problem. A scientist is performing some experiments on pure oxygen gas which will have to be repeated several times at widely different occasions. What must he know about the samples of oxygen on which the measurements are made in order to be sure that these samples are in precisely the same condition or state for each measurement? What is the minimum number of properties of pure oxygen that must be specified in order to fix completely the condition of the oxygen?

Experience tells us that under ordinary conditions this minimum number is precisely three.[2] That is, if any three of the four

[2] By "ordinary" conditions here it is meant that no extraneous influences such as electric or magnetic fields are present, and that the gravitational field acting on the sample under study is negligibly small. If these conditions are not satisfied, then additional variables must be specified beyond the minimum number of three in order to specify the state of the sample. It is also important to note that here we assume that we are dealing with substances whose composition cannot be altered. If the composition of a system is permitted to vary, then additional variables must, of course, be specified in order to fix the condition of the system, one new variable being required for each new component added to the system under study. The role of these "composition variables" will be considered presently. The composition variables are essential to the treatment of any system in which chemical reactions occur.

primary properties, mass, volume, pressure, and temperature, are specified for any substance, then the magnitude of the fourth primary property, as well as the values of all other properties of that substance, are completely determined. Thus if one fills a vessel having a volume of 100 ml at a temperature of 30°C with 32 g of oxygen, then the pressure inside the vessel will be found to be 239 atm. It is important to realize that two properties are not sufficient for this; the statement that one has 32 g of oxygen at 30°C is not sufficient to specify the state of the oxygen because any pressure between zero and infinity is consistent with these two properties, depending on the volume chosen for the container, which may have any value from infinity to zero.

The very fundamental observation concerning the nature of matter that has just been presented may be expressed in mathematical language by the statement that the properties of a sample of any given substance are functions of three independent variables. If we agree to restrict our choice of independent variables to the set of the four "primary properties," pressure, P, volume, V, temperature, T, and mass, m (an arbitrary restriction which we make for reasons of convenience alone since these properties are particularly easy to measure), then we may say that there exist for any sample of a given substance four functions: $P(V, T, m)$, $V(P, T, m)$, $T(P, V, m)$, and $m(P, T, V)$. These four functions are, of course, related to one another, in the sense that if one function is known, then the other three functions can, at least in principle, be evaluated. Furthermore, it is clear that functions must exist which give the numerical values of all other properties of any given sample of a substance in terms of any three of the four primary variables. For instance, if E is the energy content of oxygen, then functions such as $E(P, V, T)$ and $E(V, T, m)$ must exist.

An equation which relates the four primary properties, P, V, T, and m, for a substance is called the *equation of state* of that substance. We shall now consider the equations of state of gases that have been observed by physical chemists.[3] The general procedure that is followed in determining empirically the form of these equations is to keep two of the primary properties constant while varying the third, and observing the resultant behavior of the fourth property.

[3] One can also speak of equations of state for liquids and solids as well as for gases.

1-3 The ideal gas laws

If the pressure, volume, and temperature of a gas are measured at low pressures (say, below a few atmospheres), they are found to obey quite closely the so-called ideal gas law,

$$PV = nRT \tag{1-1}$$

where n is the number of moles of gas occupying a volume V at an absolute temperature T and pressure P, and R is a universal constant called the gas constant. As will now be shown, this concise law is a consequence of four generalizations discovered over a period of nearly two hundred years, known as Boyle's law, Gay-Lussac's law, Avogadro's law, and Dalton's law.

a. BOYLE'S LAW In 1662 Robert Boyle described the following simple experiment. He prepared a tube in the form of a J, the short side of the J being sealed off (Fig. 1-2; this is prototype of the manometer, mentioned in Section 1-1, above). Mercury was added through the open end, trapping some gas in the closed end. Since the tube was uniform in cross section the volume of the gas trapped in the J is proportional to the length of the column of gas, l. Part of the pressure exerted on this gas was proportional to the distance, h, between the menisci of the mercury in the two arms of the J. It was also necessary to add the pressure exerted by the earth's atmosphere on the meniscus at the open end of the tube. This pressure was proportional to h_0, the height of a mercury column in a mercury barometer, measured at the time of the experiment. Boyle observed that as mercury was poured into the open end of the J-tube, the volume, as measured by l, decreased in inverse proportion to the pressure, as measured by $h + h_0$. That is,

$$l(h + h_0) = \text{const} \tag{1-2}$$

The accuracy of this equation in expressing the results was quite good; it was possible for Boyle to predict $h + h_0$ to better than 1% for a fourfold variation in l (h being varied from 0 to $3h_0$). The experiment was performed in a room whose temperature was essentially constant, and the law is valid only under conditions of constant temperature. This result, now known as Boyle's law, can be written in the form

$$PV = \text{const} \qquad (m \text{ and } T \text{ const}) \tag{1-3}$$

EQUATIONS OF STATE OF GASES 11

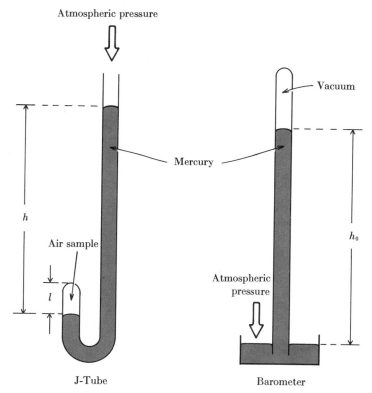

FIG. 1-2 *Boyle's J-tube experiment.*

where P and V are the pressure and volume of the gas sample. Boyle's law can be written more concisely in the form

$$PV = A(m, T) \tag{1-4}$$

where $A(m, T)$ is a quantity whose magnitude is fixed when the mass and temperature of the gas is fixed, but which does not depend explicitly on the pressure and volume of the gas. The quantity $A(m, T)$ also depends on the nature of the gas, and would be expected, for given m and T, to have different values for, say, air and carbon dioxide.

Boyle's experiment played a large part in the development of modern concepts of gases because it showed that by lowering the pressure sufficiently, one could approach the state of a vacuum—a notion that appeared to contradict the Aristotelian doctrine that "nature abhors a vacuum."[4]

b. GAY-LUSSAC'S LAW In 1802, Gay-Lussac showed [5] that if V_f is the volume of a gas at the temperature of the freezing point of water and at a pressure of one atmosphere, and if V_b is the volume of the same gas at the temperature of the boiling point of water and at the same pressure, then

$$\frac{V_b}{V_f} = K \qquad (P \text{ and } m \text{ const}) \tag{1-5}$$

where the constant K was the same for all of the gases studied. Gay-Lussac's measurements indicated a value of about 1.375 for the constant K, but more precise measurements reveal that the ratio differs slightly for different gases. If the pressure used in making the measurements is sufficiently low (well below one atmosphere) careful observation indicates that the ratio does approach a universal value. This value is 1.36609 for all gases.

Gay-Lussac's observation suggests that the thermal expansion of gases can provide a logical basis for establishing a temperature scale. Ordinary thermometers, such as those described earlier in this chapter, depend on the thermal expansion of some liquid, but all liquids do not expand in exactly the same way with temperature so the temperatures given by different liquid thermometers do not agree precisely. Since all gases expand in the same way when the temperature is changed, the pressure being held constant at a sufficiently low value, a thermometer based on gas expansion should give the same reading regardless of the gas employed.

[4] A fascinating and detailed account of Boyle's experiments and their historical significance in the development of our concepts of gases may be found in *Harvard Case Histories in Experimental Science* (J. B. Conant, ed., Vol. I, Case 1, Harvard University Press, Cambridge, Mass., 1957).

[5] According to Gay-Lussac, this law was observed by Charles in 1787, but Charles had never published his observations. The law is therefore often called Charles' law. The Italian physicist Volta published a paper containing this law in 1793. The law was also stated by Priestley in 1790 and by Dalton in 1801, but Gay-Lussac's experiments were more precise and it was he who brought the observation to general notice.

EQUATIONS OF STATE OF GASES

Therefore we may define the temperature scale by the relation

$$V = V_0(1 + \alpha t) \tag{1-6}$$

where V is the volume at temperature t, V_0 is the volume when $t = 0$ and α is related to the constant K in Eq. (1-5). Two temperature scales are in common use—the Fahrenheit scale and the Celsius scale (also called the centrigrade scale—a practice now discouraged by international agreement). For the Farenheit scale the freezing point of water at a pressure of one atmosphere is given the value $t = 32°$ and the boiling point of water at one atmosphere is given the value $t = 212°$. Setting

$$\frac{V_{212}}{V_{32}} = \frac{1 + 212\alpha}{1 + 32\alpha} = 1.36609$$

we find that the constant α has the value $1/491.7$. For the Celsius scale the freezing point of water is given the value $t = 0°$ and the boiling point of water is set at $t = 100°$, which gives α the value $1/273.16$.

For a great many applications in physical chemistry it is useful to define still another scale of temperature, called the absolute scale, defined by

$$T = \left(\frac{1 + \alpha t}{\alpha}\right) = t + \left(\frac{1}{\alpha}\right) \tag{1-7}$$

in which the temperature $T = 0$ (corresponding to $t = -1/\alpha$) corresponds to the temperature at which the volume of a gas would vanish if the gas obeyed Eq. (1-6) at low temperatures. (All gases become liquids at temperatures well above $t = -1/\alpha$, so that gas thermometers will not be useful at very low temperatures. By using helium, we can hope to use the gas thermometer down to about $-269°C$, which is the boiling point of helium.) The absolute temperature based on the value of the constant α derived from the Fahrenheit scale is called the Rankine temperature scale and is indicated by the symbol °R. Using the value of α derived from the Celsius scale, we obtain the Kelvin temperature scale, indicated by the symbol °K. We may note that [6]

$$0°C = 273.16°K = 491.7°R$$

$$0°F = 459.7°R = -17.78°C = 255.38°K$$

The Rankine scale is widely used by engineers in England and the United States, but the Kelvin scale is almost universally used in scientific work.

If Eq. (1-7) is used to replace t by T in Eq. (1-6), Gay-Lussac's law takes the particularly simple form

$$V = \alpha V_0 T$$

or

$$\frac{V}{T} = \text{const} \qquad (P \text{ and } m \text{ const}) \tag{1-8}$$

or

$$\frac{V}{T} = B(m,P) \tag{1-9}$$

where $B(m, P)$ is a quantity whose magnitude is fixed when m and P are specified for a given substance. Equation (1-9) may be taken as the mathematical expression of the law of Gay-Lussac.

c. CONSOLIDATION OF BOYLE'S LAW WITH GAY-LUSSAC'S LAW The behavior of a sample of gas at fixed temperature is given by Eq. (1-4) whereas Eq. (1-9) describes the behavior of the sample when the pressure is fixed. It is desirable to combine these two equations into a single equation which will describe the behavior of a gas sample when both P and T are varied simultaneously. This may be accomplished in the following manner. From Eq. (1-4) we have

$$V = \frac{A(m, T)}{P}$$

[6] Recently the Kelvin scale of temperatures has been redefined so that instead of using the ice point and the boiling point of water as fixed points, 0°K is defined as one of the fixed points and the triple point of water (i.e., the temperature at which ice, liquid water, and water vapor are in equilibrium at the vapor pressure of the water) is defined as 273.16°K. The ice point (i.e., the melting point of ice under a pressure of one atmosphere) is 0.0100 ± 0.0001° below the triple point. Thus by international agreement 0°C is now given the value 273.15°K, and a temperature interval of 1°C is no longer precisely equal to a temperature interval of 1°K.

EQUATIONS OF STATE OF GASES

and from Eq. (1-9) we have

$$V = B(m, P)T$$

Combining these two equations we find that

$$\frac{A(m, T)}{P} = B(m, P)T$$

or

$$\frac{A(m, T)}{T} = B(m, P)P \tag{1-10}$$

This equation must, of course, be true for all values of the pressure and temperature. Clearly, when the temperature is varied while the pressure and mass are held constant, the right-hand side of Eq. (1-10) must remain fixed. Consequently, the quantity $A(m, T)/T$ on the left-hand side of Eq. (1-10) must also be independent of the temperature. Thus we may write

$$\frac{A(m, T)}{T} = C(m)$$

where $C(m)$ is a function whose numerical value depends on the quantity of gas present, but not on P, T, or V. (For a given value of m we must also expect that the magnitude of C will be different for different gases.) Thus we have deduced the nature of the temperature dependence of the function $A(m, T)$ in Eq. (1-4),

$$A(m, T) = C(m)T \tag{1-11}$$

On inserting this relation into Boyle's law, Eq. (1-4), we find

$$V = \frac{C(m)T}{P} \tag{1-12}$$

which enables us to calculate, say, the volume of a given quantity of gas when both the pressure and the temperature are changed.

It is an easy matter to deduce the dependence of the function $C(m)$ on the quantity of gas present, m. Our experience tells us that if we increase the quantity of a given kind of gas by some factor, holding the temperature and pressure fixed, the volume of the gas will be increased by the same factor. (For instance, 3 g of oxygen at 1 atm and 30°C occupies three times the volume of 1 g of oxygen at the same pressure and temperature.) Thus we must conclude that $C(m)$ is linear in m, so that

$$V = \frac{mDT}{P} \tag{1-13}$$

where D is a constant which is independent of m, P, and T, but which has different values for different kinds of gases.

d. AVOGADRO'S LAW In 1811 Avogadro proposed that at a given temperature and pressure, equal volumes of all gases contain equal numbers of molecules. His reasons for making this hypothesis were complex, and it was not generally accepted until 1858, when it was revived by Cannizzaro.[7] The hypothesis is now fully accepted and deserves to be recognized as a law, characterizing the properties of gases with a precision equivalent to that attained by the laws of Boyle and of Gay-Lussac. Avogadro's law can be expressed by the equation

$$V = nV_0(P, T) \tag{1-14}$$

where V_0 is the same function for all gases and n is some measure of the number of molecules of gas present.

In scientific work the quantity of a substance whose mass, in grams, is equal to the molecular weight is called a "gram-mole," or more usually merely a "mole."[8] The number of molecules present in one mole of a substance is the same for all substances. Thus the number of moles of gas in a system is a convenient measure of the number of molecules present. We can therefore write

$$n = \frac{m}{M} \tag{1-15}$$

and

$$V = \frac{V_0(P, T)m}{M} \tag{1-16}$$

where m is the mass of the substance and M is its molecular

---

[7] The interesting history of this idea is given by L. K. Nash in *Harvard Case Histories in Experimental Science* (J. B. Conant, ed., Vol. I, Case 4, Harvard University Press, Cambridge, Mass., 1957).

[8] In chemical engineering it is sometimes convenient to use the concept of the "pound-mole," which is the mass in pounds equal to the molecular weight (one pound-mole of oxygen contains 32 lb of oxygen, whereas one gram-mole of oxygen contains 32 g).

EQUATIONS OF STATE OF GASES

weight. If m is expressed in grams, then n is the number of gram-moles present whereas if m is expressed in pounds, then n is the number of pound-moles.

e. THE IDEAL GAS LAW The volume V may be eliminated from Eqs. (1-13) and (1-16) to give

$$V_0(P, T) = \frac{DMT}{P} \tag{1-17}$$

Since V_0 is the same function for all gases, this relation signifies that the quantity DM must be a universal constant. This constant is called the *gas constant* and is conventionally indicated by the symbol R

$$R = DM$$

so that

$$V_0 = \frac{RT}{P}$$

and

$$V = \frac{m}{M}\frac{RT}{P} = \frac{nRT}{P}$$

or

$$PV = \frac{mRT}{M} = nRT \tag{1-18}$$

which is the ideal gas law, expressing in one relationship the laws of Boyle, Gay-Lussac, and Avogadro. It is an excellent approximation to the equation of state of real gases at low pressures, and may be called the *equation of state of an ideal gas*. The circumstances under which it fails to describe the behavior of real gases, and the magnitude of this failure, will be discussed below.

f. UNITS OF MEASUREMENT OF THE PRIMARY VARIABLES AND THE NUMERICAL VALUE OF THE GAS CONSTANT Some of the units in which the primary variables are commonly measured are listed in Table 1-1, along with conversion factors relating the different systems of units to each other. Most of these units are undoubtedly already familiar to the student, but the use of mercury columns to measure pressure deserves some special discussion. If a tube is closed at one end, filled with mercury and inverted

TABLE 1-1 UNITS FOR THE MEASUREMENT OF THE PRIMARY VARIABLES

Variable	Unit

Volume	cgs unit = 1 cm^3 = 0.99997 ml
	 1 cubic inch = 16.387 ml
	 1 cubic foot = 28,316 ml
	 1 liter = 1000 ml = 1000.027 cm^3
	         = 61.025 in.^{3}
	         = 0.035316 ft^3

Mass	cgs unit = 1 g 1 pound = 453.59237 g 1 pound mole = 453.59237 gram moles

Pressure	cgs unit = 1 dyne/cm^2 = 1 barye
	 1 bar = 10^6 baryes
	 1 millibar = 10^3 baryes
	 1 normal atm = 1,013,250 baryes
	               = 76 cm Hg (0°C, g = 980.665 cm/sec^2)
	               = 1.0332 kg/cm^2
	               = 14.696 lb/in.^{2}
	 1 torr = 1 mm Hg (0°C, 980.665 cm/sec^2)
	 1 μ = 0.001 mm Hg (0°C, 980.665 cm/sec^2)

Temperature	t°C = $(9t/5 + 32)$°F 0°C = 273.15°K = 491.7°R 0°F = 459.7°R

into a dish of mercury in a container filled with a gas, it is known that the gas is able to support a column of mercury in the tube whose length is a convenient measure of the pressure of the gas. The pressure of the gas is equal to the force exerted by a unit cross section of the mercury at the base of the column. This pressure is given by

$$P = \rho g h \qquad (1\text{-}19)$$

where ρ is the density of the mercury in the column, g is the

EQUATIONS OF STATE OF GASES

acceleration due to gravity, and h is the length of the column. Now, the density of mercury in this equation depends on the temperature of the mercury in the column. Furthermore the magnitude of g varies with one's location on the earth. Therefore, in order to be able to express P in terms of h, there must be some convention as to the numerical values of ρ and g to be used in Eq. (1-19). The internationally accepted definition of 1 cm of mercury as a unit of pressure specifies that the temperature of the mercury be 0°C and the value of g be 980.665 cm/sec². If mercury columns are used under other conditions, appropriate corrections must be made.

EXERCISE The barometric pressure is measured in New York City by means of a mercury barometer in a room at 68°F. The column is found to be 30.124 in. long. Find the pressure in normal atmospheres. (Obtain values of the density of mercury and g in New York City from a handbook.)

The numerical value of the gas constant, R, will depend on the units used in expressing the four primary variables in the equation of state. Values of R for certain important combinations of these units are given in Table 1-2. It is interesting to note that the product PV has the dimensions of energy. This is evident from the dimensional relations

$$\begin{aligned}
[\text{energy}] &= [\text{force}] \times [\text{distance}] \\
&= [\text{mass}] \times [\text{acceleration}] \times [\text{distance}] \\
&= M \times [LT^{-2}] \times L \\
&= ML^2T^{-2}
\end{aligned}$$

$$\begin{aligned}
[\text{pressure}] \times [\text{volume}] &= [\text{force/area}] \times [\text{volume}] \\
&= [MLT^{-2}] \times L^{-2} \times L^3 \\
&= ML^2T^{-2}
\end{aligned}$$

Thus the gas constant, R, has the dimensions [energy/temperature × moles]. Later in this book it will be convenient to express the magnitude of R in units such as (joules/°K gram moles) and (calories/°K gram moles), despite the fact that these units do not correspond to any combination of pressure and

TABLE 1-2 NUMERICAL VALUES OF THE GAS CONSTANT, R

Units employed for the primary variables

Pressure	Volume	Temperature	Mass	R
Atmospheres	liters	°K	g moles	0.0820561
Atmospheres	ml	°K	g moles	82.0561
Baryes	cm³	°K	g moles	8.31433×10^7
cm Hg	ml	°K	g moles	6,236.26
kg/cm²	ml	°K	g moles	84.780
Atmospheres	in.³	°R	lb moles	1,261.8
Atmospheres	ft³	°R	lb moles	0.7302
lb/in.²	in.³	°R	lb moles	18,544.
lb/in.²	ft³	°R	lb moles	10.732
R in joules (°K)$^{-1}$(g mole)$^{-1}$				8.31433[a]
R in calories (°K)$^{-1}$(g mole)$^{-1}$				1.98717[a]

[a] From F. D. Rossini, *Pure Appl. Chem.*, **9**, 453 (1964).

volume units that are in practical use. The values of R expressed in some of these energy units are also included in Table 1-2.

g. DALTON'S LAW The ideal equation of state, Eq. (1-18), has been developed assuming that a single substance is present in the system under study. A law discovered by Dalton makes it an easy matter to extend the ideal gas law to mixtures of several substances. It is well known that all gases are completely soluble in one another. Dalton observed that when several gases, which do not react chemically, are mixed together in a given volume, the pressure that results is the sum of the pressures that would be observed if each gas occupied the volume by itself (*Dalton's law*).

If a container of volume V contains n_1 moles of a gas 1, n_2 moles of gas 2, etc., then we can define the *partial pressures* of gases 1, 2, ... , at temperature T by the relations

$$p_1 = \frac{n_1 RT}{V} \qquad p_2 = \frac{n_2 RT}{V} \qquad \cdots \qquad (1\text{-}20)$$

EQUATIONS OF STATE OF GASES

The partial pressure of a gas is therefore the pressure that the gas would exert if it were present by itself in the container. According to Dalton's law, the total pressure of the mixture is

$$P = p_1 + p_2 + \cdots = \sum_{\text{all constituents, } i} p_i \qquad (1\text{-}21)$$

This can also be written as

$$P = \frac{\sum_i n_i RT}{V} = \frac{nRT}{V} \qquad (1\text{-}22)$$

where n is the total number of moles of gas present,

$$n = n_1 + n_2 + \cdots = \Sigma n_i \qquad (1\text{-}23)$$

Equation (1-22) is an even more general form of the ideal gas law, valid for mixtures as well as for pure gases; it combines in one equation the four laws of Boyle, Gay-Lussac, Avogadro, and Dalton.

It is convenient to define the *mole fractions*, x_i, of the various constituents of the mixture ($i = 1, 2, \ldots$) by means of the expression

$$x_i = \frac{n_i}{n} \qquad (1\text{-}24)$$

The partial pressure p_i of constituent i is then given by the product of the mole fraction and the total pressure

$$p_i = x_i p \qquad (1\text{-}25)$$

Note that

$$\sum_{\text{all constituents, } i} x_i = 1 \qquad (1\text{-}26)$$

In a mixture the quantities x_i may be considered to be additional variables, required along with the four primary variables in order to specify the state of the mixture. Because of Eq. (1-26) there must be one fewer composition variables than there are components in the mixture. (Obviously for a system containing only one component there can be no composition variables.)

1-4 The equations of state of real gases

a. GENERAL NATURE OF THE DEVIATIONS OF REAL GASES FROM THE IDEAL GAS LAW The ideal gas law is obeyed with reasonably good precision by real gases at low pressures, but accurate measurements reveal the existence of deviations which become increasingly important as the pressure is raised. The nature of these deviations from ideal behavior is most clearly shown by observing the changes in the quantity $z = PV/nRT$ as the pressure is raised. The quantity z is often called the *compressibility factor* of a gas.[9] If the ideal gas law were obeyed at all pressures and all temperatures, then the compressibility factor would always be equal to unity, and a plot of z vs. P would show a horizontal line at all temperatures (Fig. 1-3). In reality the compressibility factor shows a relatively complex behavior when the temperature and pressure are changed.

At low pressures (that is, below a few atmospheres for the common gases such as air, hydrogen, helium, etc., at ordinary temperatures) it is observed that a plot of z vs. P gives a straight line which extrapolates to $z = 1$ when $p = 0$. The slope of this line varies widely from one gas to another (Fig. 1-3), and for a given gas the slope changes considerably with the temperature (see Fig. 1-4). For different gases at the same temperature there is a definite correlation between the slope and the boiling point of the gas; substances with higher boiling points (such as CO_2 and C_2H_4 in Fig. 1-3) tend to give more negative slopes, whereas substances with low boiling points (such as Ne and H_2 in Fig. 1-3) tend to give less negative or even slightly positive slopes. For a given gas it is found that the slope of the z vs. P plot becomes more negative as the temperature is lowered (Fig. 1-4).

At higher pressures (of the order of hundreds of atmospheres or more) the plots of z vs. P take on a more complicated appearance (Fig. 1-5). At room temperature these plots go through a minimum at pressures of approximately 100 to 200 atm for many of the common gases. The minimum is more pronounced, and occurs at lower pressures, the higher the boiling point of the gas.

---

[9] The *compressibility factor*, $z = PV/nRT$, should not be confused with an entirely different property, called the *compressibility*, which is defined as $-(1/V)(dV/dP)$, and which measures the ease with which an object's volume can be decreased when pressure is applied.

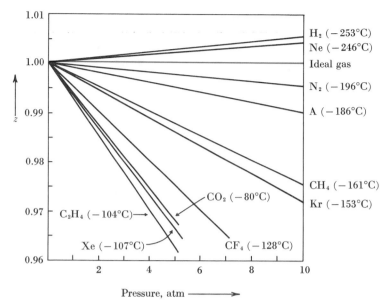

FIG. 1-3 *Variation of the compressibility factor, $z = PV/nRT$, with pressure at $0°C$ for various gases. The boiling points of the gases are indicated in parentheses.*

This behavior of z as a function of P can be expressed conveniently in the form of a polynomial, known as the *virial equation*,

$$z = \frac{PV}{nRT} = 1 + B'(T)P + C'(T)P^2 + D'(T)P^3 + \cdots$$

(1-27)

The coefficients $B'(T)$, $C'(T)$, $D'(T)$, ..., are known as the second, third, fourth, ..., virial coefficients, respectively. The prime is used here because later we shall introduce another form of the virial equation involving a polynomial in n/V, whose coefficients must be distinguished from the present ones. Note that the "first virial coefficient"—that is, the first term of the polynomial—is unity because the ideal gas law becomes more nearly correct as the pressure is reduced to zero. The numerical values of the coefficients depend upon the temperature of the gas, so they are written as functions of T.

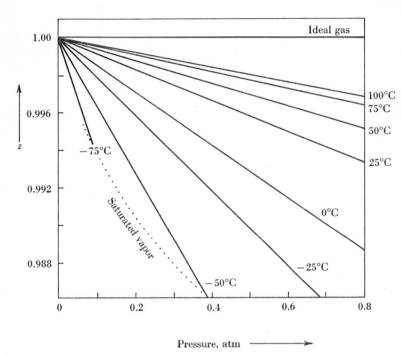

FIG. 1-4 *Dependence of the compressibility factor of ammonia on the pressure at various temperatures. The dotted line gives the locus of the compressibility factors at the saturated vapor pressures of ammonia.*

At low pressures the terms $C'P^2$, $D'P^3$, etc., can be neglected in comparison with the terms unity and $B'P$, and we can write

$$\frac{PV}{nRT} = 1 + B'(T)P \tag{1-28}$$

which expresses the result shown in Figs. 1-3 and 1-4. The slopes of the lines in these figures are the second virial coefficients. We may say that $B'(T)$ usually tends to be negative, and that it becomes more negative as the temperature is lowered. It was observed empirically by Berthelot that to a fairly good approximation

$$B'(T) = \frac{a}{T} - \frac{b}{T^n} \tag{1-29}$$

where a, b, and n are positive constants. The constant n has a value close to 3 for many substances. Therefore the second term on the right in Eq. (1-29) determines the sign of B' at low temperatures, and the first term determines the sign at high temperatures.

We shall find in the next chapter that second virial coefficients can tell us a great deal about the forces acting between molecules in a gas.

b. THE CRITICAL POINT OF A GAS   Another way of viewing the behavior of real gases is to plot the P-V isotherms (that is, the

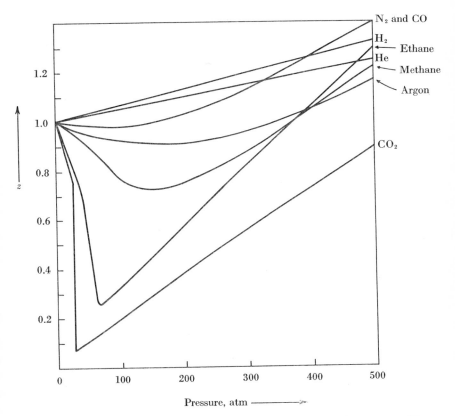

FIG. 1-5 *Variation of the compressibility factor of various gases with pressure at high pressures. The temperature is 0°C for all gases except ethane, for which the temperature is 37°C.*

pressure vs. volume graphs at a series of constant temperatures) and to compare the result with the ideal gas law. Let us assume that one mole of a substance is introduced into a cylindrical container one end of which is fitted with a movable piston, the device being then immersed in a bath whose temperature can be held constant (Fig. 1-6a). The volume of the system is varied by moving the piston, and the variation of the pressure of the gas acting on the piston is observed. The ideal gas law predicts that $P = RT/V$. This relationship is plotted in Figs. 1-6b and 1-6c. Figure 1-6b shows the variation of P with V at a series of bath temperatures as a family of curves in two dimensions whereas Fig. 1-6c shows the relationship of P, V, and T as a surface in three dimensions. What is experimentally observed in a typical gas, say ammonia, in the temperature range up to 150°C and at pressures up to a few hundred atmospheres is shown in Fig. 1-6d as a family of isotherms and in Fig. 1-6e as a three-dimensional plot. Perhaps the most striking difference between the observed behavior and that predicted by the ideal gas law is found below a temperature of 132.5°C, where the pressure remains constant over a finite range of volumes. For instance, at 100°C the pressure is constant at 61.8 atm if the mole volume lies between 37 and 300 ml. If the container were fitted with a window, so that we could watch what was happening to the gas under these conditions, we would find that two layers of ammonia were present in the container—a light layer which corresponds to gaseous ammonia and a denser layer which we would recognize as liquid ammonia. The two layers would be separated by a sharp boundary, or meniscus. When, at 100°C, the volume per mole is decreased from 300 to 37 ml, the amount of gas present decreases and the amount of liquid increases. At a volume of 300 ml or more the ammonia is entirely gaseous, and at any volume less than 37 ml it is entirely liquid. The constant pressure at which this change takes place at 100°C is called the *vapor pressure* at that temperature.

At any temperature below 132.5°C the volume at the left end of the constant pressure segment of the isotherm is the mole volume of liquid ammonia and the volume at the other end is the mole volume of the gas that is in equilibrium with the liquid at that temperature. Note that the P-V isotherm undergoes discontinuities in slope at the two ends of the constant pressure segment.

As the temperature of the system is varied, the two ends of the constant pressure line segment trace out loci that are indicated by colored lines in Figs. 1-6d and 1-6e. It is evident that the horizontal line segments become shorter as the temperature is raised. The reason for this is as follows. The vapor pressure increases rapidly as the temperature is raised. The density of the vapor that is in equilibrium with the liquid therefore increases, and the mole volume of this vapor decreases with increasing temperature. The right end of the constant pressure segment of the P-V isotherm therefore moves toward the left as the temperature is raised. On the other hand, the liquid expands as the temperature is raised, so that the left end of the constant pressure segment moves toward the right. Thus the two ends of the segment move toward each other on raising the temperature, so that the segment becomes shorter.

It is evident from Figs. 1-6d and 1-6e that at a temperature of 132.5°C the mole volumes of ammonia liquid and vapor have become identical; the colored line corresponding to liquid ammonia meets the colored line corresponding to ammonia vapor at the point C. This point C on the P-V isotherm is called the *critical point*. The temperature at which the mole volumes of the liquid and vapor become equal is called the *critical temperature*. The pressure at the critical point is called the *critical pressure* (P_c in Figs. 1-6d and 1-6e) and the volume for this condition is called the *critical volume* (V_c in Figs. 1-6d and 1-6e).

Above the critical temperature there is no clear distinction between the liquid and vapor (or gaseous) states of a substance because the two states cannot coexist with a sharp boundary between them. If one mole of a gas is placed in a transparent tube whose volume is equal to the critical volume, and if the tube is held below the critical temperature, two layers will be observed, separated by a sharp boundary. As the system is warmed the boundary becomes less distinct because the densities, and therefore the refractive indices, of the liquid and vapor approach a common value. When the critical temperature is reached, the boundary becomes invisible. This striking experiment is not very difficult to perform and offers a means of determining the critical temperature with considerable precision.

It is interesting to note the possibility of transforming a liquid into a gas without at any time observing a sharp boundary separating the two states. This may be accomplished (see Fig. 1-6e)

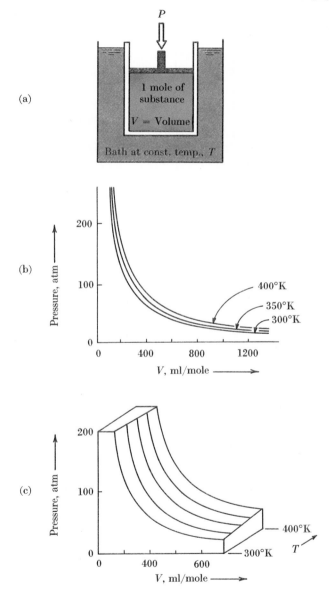

FIG. 1-6 *The P-V isotherms of ideal and real gases. (a) Conceptual device for measuring the isotherms of a gas or liquid. (b) Ideal gas isotherms for 300° to 400°K and pressures to 200 atm. Two-dimensional representation. (c) Three-dimensional representation of the equation of state of an ideal gas between 300°K and 400°K.*

EQUATIONS OF STATE OF GASES

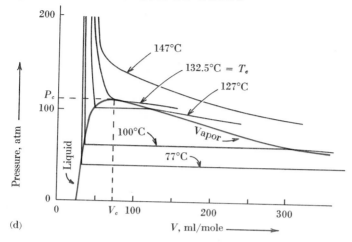

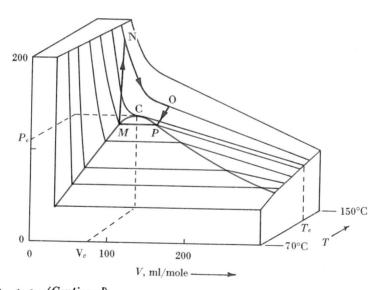

FIG. 1-6 (*Continued*)
(*d*) Two-dimensional representation of the observed isotherms of ammonia at 0 to 200 atm. and 77°C to 147°C. The point C is the critical point.
(*e*) Three-dimensional representation of the equation of state of ammonia at 70° to 150°C and 0 to 200 atm.

by first heating the liquid to a temperature above the critical temperature, while maintaining the volume constant (process $M \to N$ in Fig. 1-6e). The volume is then increased at constant temperature until it reaches a value corresponding to that of the gas at the original temperature (process $N \to O$ in Fig. 1-6e). The temperature is then reduced to its original value, the volume being held constant (process $O \to P$ in Fig. 1-6e). The substance will now be in the gaseous state, but at no stage in the process $M \to N \to O \to P$ can it be said that the liquid and vapor coexisted as two distinct phases.

Values of the critical constants of a number of substances are given in Table 1-3. Also given in this table are the ratios $P_c V_c / RT_c$, T_c/T_b, and V_c/V_b, where T_b is the boiling point at one atmosphere and V_b is the mole volume at the boiling point. It is seen that the ratio T_c/T_b tends to have a value close to 1.6, V_c/V_b has a value close to 2.7, and $P_c V_c / RT_c$ is usually in the range 0.25–0.30. It is much easier to measure the boiling point and the mole volume at the boiling point than it is to measure T_c, V_c, and P_c because critical pressures are generally quite high and special pressure equipment is needed to work at these pressures. These empirical ratios therefore make it possible to estimate the magnitudes of the critical constants of substances for which they have not been measured directly.

c. EQUATIONS OF STATE OF REAL GASES The equation of state of a real gas must be complex, as is evident from Figs. 1-5 and 1-6; it must become particularly complicated when the gas undergoes a transformation to the liquid state, causing discontinuities to occur in the slopes of the P-V isotherms. A considerable effort has been made to develop theoretical or empirical equations of state which represent the behavior of real gases.

The most general of these equations are the virial equations of state, in one form of which the compressibility factor is expanded in powers of the pressure,

$$\frac{PV}{nRT} = 1 + B'P + C'P^2 + D'P^3 + \cdots \tag{1-30}$$

the coefficients B', C', D', ... being functions of the temperature. The observed temperature dependence of B' has already been discussed. An equivalent form of the virial equation which is

EQUATIONS OF STATE OF GASES

TABLE 1-3   CRITICAL CONSTANTS OF GASES

Gas	T_c (°K)	P_c (atm)	V_c (ml/mole)	$\dfrac{T_c}{T_b}$	$\dfrac{V_c}{V_b}$	$\dfrac{P_c V_c}{RT_c}$
He	5.3	2.26	57.7	1.26	1.81	0.300
Ne	44.5	25.9	41.6	1.63	2.49	0.296
A	151.	48.0	75.2	1.72	2.64	0.291
Kr	210.6	54.2	92.	1.76		0.290
Xe	290.	57.9	120.2	1.77		0.293
Rn	377.	62.4		1.74		
H_2	33.2	12.8	65.0	1.64	2.30	0.304
N_2	126.0	33.5	90.0	1.63	2.6	0.292
O_2	154.3	49.7	74.4	1.71	2.65	0.292
Cl_2	417.	76.1	123.	1.75	2.73	0.274
Br_2	575.			1.73		
I_2	826.			1.81		
NO	183.	65.	58.	1.50	2.44	0.251
CO	134.	35.	90.	1.70	2.61	0.287
CO_2	304.2	73.0	95.6	1.56		0.287
CS_2	546.	75.		1.71		
N_2O	310.	71.7	97.5	1.69	2.72	0.274
N_2O_4	431.	99.		1.47		
SO_2	430.	77.7	123.	1.59		0.271
SO_3	491.	83.6	127.	1.54	2.87	0.263
O_3	268.	66.	89.4	1.66		0.268
HCl	325.	81.6	86.2	1.73	2.81	0.264
HBr	383.	85.		1.86		
HI	424.	82.		1.78		
HCN	457.	50.	135.	1.53		0.180
H_2O	647.1	217.7	45.0	1.735	2.41	0.184
D_2O	644.5			1.72		
H_2S	373.5	88.9		1.75		
NH_3	405.5	111.5	72.3	1.69	2.91	0.238
PH_3	324.	64.	110.	1.74	2.5	0.265
CH_4	190.6	45.8	98.8	1.63	2.56	0.290
C_2H_6	305.2	49.	136.	1.65	2.49	0.267
C_2H_4	282.9	50.9	127.5	1.67		0.291
C_2H_2	309.	62.	60.1			0.275
C_3H_8	369.9	45.3	181.	1.62	2.42	0.270

TABLE 1-3 CRITICAL CONSTANTS OF GASES Continued

Gas	T_c (°K)	P_c (atm)	V_c (ml/mole)	$\dfrac{T_c}{T_b}$	$\dfrac{V_c}{V_b}$	$\dfrac{P_c V_c}{RT_c}$
n-C$_4$H$_{10}$	426.0	42.0	214.	1.56	2.21	0.257
i-C$_4$H$_{10}$	407.1	37.	248.	1.55	2.58	0.276
C$_6$H$_6$	561.6	47.7	257.	1.59		0.268
Methyl alcohol	513.1	78.7	117.5	1.52	2.76	0.221
Ethyl alcohol	516.	63.1	167.	1.47	2.68	0.250
n-Propyl alcohol	537.	50.	219.	1.45	2.69	0.248
Methyl amine	330.	73.6		1.25		
Acetic acid	594.	57.2	171.	1.52	2.68	0.201
Acetonitrile	547.8	47.7	171.	1.54		0.181
Methyl chloride	416.1	65.8	136.5	1.66		0.258
Ethyl chloride	460.3	53.	195.	1.61	2.76	0.263
Ethyl acetate	523.	37.8	286.	1.49	2.69	0.253
Dimethyl ether	400.	52.	170.	1.60	2.67	0.270
Diethyl ether	466.9	35.5	281.	1.51	2.65	0.261
Acetone	508.	47.	216.	1.54	2.70	0.243

more useful for some purposes is an expansion of the compressibility factor in reciprocal powers of the mole volume, n/V.

$$\frac{pV}{nRT} = 1 + B\frac{n}{V} + C\left(\frac{n}{V}\right)^2 + D\left(\frac{n}{V}\right)^3 + \cdots \qquad (1\text{-}31)$$

The coefficients B, C, D, ... may be related to the coefficients B', C', D', ... in the following way. From Eq. (1–30) we may write

$$\frac{n}{V} = \frac{P}{RT}(1 + B'P + C'P^2 + \cdots)^{-1} \qquad (1\text{-}32)$$

Now it is well known that, from the binomial theorem, if $x < 1$,

$$(1 + x)^{-1} = 1 - x + x^2 - \cdots \qquad (1\text{-}33)$$

Writing $x = B'P + C'P^2 + \cdots$ and disregarding terms involving powers of P greater than two (which will be satisfactory at low pressures), we find

$$(1 + B'P + C'P^2 + \cdots)^{-1} = 1 - B'P - C'P^2 + B'^2P^2 \qquad (1\text{-}34)$$

EQUATIONS OF STATE OF GASES

so that from Eq. (1-32)

$$\frac{n}{V} = \frac{P}{RT} - \frac{B'P^2}{RT} - \frac{(C' - B'^2)P^3}{RT} + \cdots \quad (1\text{-}35)$$

Substituting Eq. (1-35) into Eq. (1-31) and disregarding terms involving powers of P greater than two, we find

$$\frac{PV}{nRT} = 1 + \frac{B}{RT}P + \left[\frac{C}{R^2T^2} - \frac{B'B}{RT}\right]P^2 + \cdots \quad (1\text{-}36)$$

Equating the coefficient of P in Eq. (1-36) to the coefficient of P in Eq. (1-30), we obtain

$$B' = \frac{B}{RT} \quad \text{or} \quad B = B'RT \quad (1\text{-}37)$$

and equating the coefficient of P^2 in Eq. (1-36) to that in Eq. (1-30),

$$C' = \frac{C}{R^2T^2} - \frac{B'B}{RT} = \frac{C - B^2}{R^2T^2} \quad \text{or} \quad C = R^2T^2(C' + B'^2) \quad (1\text{-}38)$$

These two equations show the relationships between the second and third virial coefficients in Eqs. (1-30) and (1-31).

As has been mentioned, the coefficients in the virial equations are functions of the temperature, and it is desirable to know what the temperature dependence actually is. Bertholet showed that for many gases the second virial coefficient is given quite satisfactorily by

$$B = \frac{9}{128}\frac{RT_c}{P_c}\left[1 - 6\left(\frac{T_c}{T}\right)^2\right] \quad (1\text{-}39)$$

where T_c and P_c are the critical temperature and critical pressure of the gas. Relatively little is known about the temperature variation of the third virial coefficients, C and C'. It appears that at temperatures well above the critical temperature, C is positive and decreases slowly with increasing temperature. There are theoretical reasons for believing that on lowering the temperature, C should pass through a maximum at about the critical temperature and then rapidly decrease to negative values at still lower temperatures.[10]

The virial equations of state are not useful at high pressures—

[10] J. O. Hirschfelder, C. F. Curtiss, and R. B. Bird, *Molecular Theory of Gases and Liquids*, Wiley, New York, 1954, p. 171.

especially at pressures around and above the critical pressure.[11] Several other equations of state have been proposed for use under these conditions:

Van der Waals' equation of state,

$$\left(P + \frac{n^2 a}{V^2}\right)(V - nb) = nRT \tag{1-40}$$

Berthelot's equation of state,

$$\left(P + \frac{n^2 a}{TV^2}\right)(V - nb) = nRT \tag{1-41}$$

Dieterici's equation of state,

$$Pe^{na/VRT}(V - nb) = nRT \tag{1-42}$$

The van der Waals equation is particularly interesting because it was originally derived by van der Waals from simple molecular considerations, as will be shown in Chapter 2.

Each of these equations contains two empirical parameters, a and b, in addition to the gas constant, R. At low pressures (large volumes) they all reduce to the ideal gas law. It is interesting to rewrite them in forms resembling the virial equation.

When the van der Waals equation is multiplied out and rearranged we obtain

$$\frac{PV}{nRT} = \frac{1 - (a/RT)(n/V) + (ab/RT)(n/V)^2}{1 - b(n/V)} \tag{1-43}$$

The factor $[1 - b(n/V)]^{-1}$ can be expanded using Eq. (1-33) giving

$$\frac{PV}{nRT} = \left[1 - \frac{a}{RT}\frac{n}{V} + \frac{ab}{RT}\left(\frac{n}{V}\right)^2\right]\left[1 + b\frac{n}{V} + b^2\left(\frac{n}{V}\right)^2 + \cdots\right] \tag{1-44}$$

[11] The reason for this is as follows. It is difficult to measure accurately the values of the third virial coefficients, C and C', and even approximate values of the higher coefficients, D, D', ... are very difficult to obtain. At pressures near to and greater than the critical pressure, however, the terms $C(n/V)^2$, $C'P^2$, $D(n/V)^3$, $D'P^3$, ... become comparable with unity and with the terms $B(n/V)$ and $B'P$. Thus the virial equation converges slowly and the higher terms are not accurately known. The problem of evaluating meaningful virial coefficients from experimental data is discussed by A. Michels, J. C. Abels, C. A. Ten Sedam, and W. de Graaff, *Physica*, **26**, 381 (1960).

which on multiplying out and retaining only terms involving the first and second powers of (n/V) gives

$$\frac{PV}{nRT} = 1 + \left(b - \frac{a}{RT}\right)\left(\frac{n}{V}\right) + b^2 \left(\frac{n}{V}\right)^2 + \cdots \qquad (1\text{-}45)$$

Thus, according to the van der Waals equation the second virial coefficients B' and B of Eqs. (1-30) and (1-31) should be given by

$$B = b - \frac{a}{RT} \quad \text{and} \quad B' = \frac{b}{RT} - \frac{a}{R^2 T^2} \qquad (1\text{-}46)$$

This agrees fairly well with what is observed, in that B and B' tend to be positive at high temperatures and increasingly negative at low temperatures (a and b are both found to be positive quantities in real gases). As we have already mentioned, however, Berthelot found that the observed temperature dependence of B' is more nearly of the form $(b/T) - (a/T^3)$. The van der Waals equation gives a positive, temperature-independent third virial coefficient,

$$C = b^2 \qquad (1\text{-}47)$$

which is probably not correct, especially at low temperatures.

A similar treatment of the Berthelot equation is easily shown to give

$$\frac{PV}{nRT} = 1 + \left[b - \frac{a}{RT^2}\right]\frac{n}{V} + b^2 \left(\frac{n}{V}\right)^2 + \cdots \qquad (1\text{-}48)$$

which leads to a temperature dependence of B and B' that is closer to the observed behavior than that predicted by the van der Waals equation. The third virial coefficient is, however, probably not correct.

The Dieterici equation can be shown to give

$$\frac{PV}{nRT} = 1 + \left(b - \frac{a}{RT}\right)\frac{n}{V}$$
$$+ \left(b^2 - \frac{ab}{RT} + \frac{a^2}{2R^2 T^2}\right)\left(\frac{n}{V}\right)^2 + \cdots \qquad (1\text{-}49)$$

giving a second virial coefficient of the same form as that obtained from the van der Waals equation. The third virial coefficient is no longer independent of the temperature, and is probably an improvement over that obtained from the other two equations; it is, however, undoubtedly incorrect at low temperatures.

Other equations of state have been proposed which contain more than two empirical parameters. Perhaps the best of these, capable of representing the P-V-T data of gases up to 200–300 atm, is the Beattie-Bridgeman equation,

$$\frac{PV}{nRT} = \left(1 - \frac{nc}{VT^3}\right)\left(1 + \frac{nB_0}{V} - \frac{n^2 bB_0}{V^2}\right) - \frac{nA_0}{RTV}\left(1 - \frac{na}{V}\right) \tag{1-50}$$

where a, A_0, b, B_0, and c are empirical constants which have been tabulated for numerous gases.[12] This equation gives for the second virial coefficient,

$$B = B_0 - \frac{A_0}{RT} - \frac{c}{T^3} \tag{1-51}$$

for the third virial coefficient,

$$C = \frac{aA_0}{RT} - \frac{cB_0}{T^3} - bB_0 \tag{1-52}$$

and for the fourth virial coefficient

$$D = \frac{bB_0 c}{T^3} \tag{1-53}$$

The remaining virial coefficients vanish in an expansion in powers of (n/V).

At the present time there is more and more interest—both practical and scientific—in the properties of gases at very high temperatures and pressures (for example, in high pressure chemical synthetic processes, in the gases produced in the burning of a propellant in a gun or rocket, and in the gases produced by a detonation wave in a solid high explosive). The equations of state of van der Waals, Berthelot, Dieterici, and Bridgeman and Beattie have all been derived from the observed behavior of gases at relatively low pressures and temperatures, and they are of only limited validity when applied under extreme conditions. Fortunately the molecular theory underlying the equation of state has been developed to such an extent that reasonably reliable guesses can often be made about the true nature of the equation under these extreme condi-

[12] H. S. Taylor and S. Glasstone, *Treatise on Physical Chemistry*, Van Nostrand, New York, 1951, Vol. 2, p. 206.

tions. This is an excellent example of the value of fundamental scientific understanding in practical situations.

d. THE LAW OF CORRESPONDING STATES The equations of state mentioned in the previous section can all be shown to account, in a certain sense, for the critical phenomenon in gases. This may be illustrated by considering the van der Waals equation, which, if we take $n = 1$, may be written in the form

$$P = \frac{RT}{V - b} - \frac{a}{V^2} \tag{1-54}$$

The student should recall that the constants a and b are always observed to be positive. If this equation is plotted on a P-V diagram for a given pair of (positive) values of a and b and for a set of values of T, a family of curves such as that shown in Fig. 1-7 is obtained. Aside from the "undulation" that is observed at low temperatures (such as the segment $ABDEF$ at temperature T_1 in Fig. 1-7) these curves resemble the P-V isotherms of real gases. It will be shown in a later volume that, because of the second law of thermodynamics, it is impossible to observe the "undulation" in real systems at equilibrium; the observed isotherm at temperature T_1 in Fig. 1-7 must be the horizontal line drawn in such a way that the areas ABD and DEF are equal. The horizontal line corresponds, of course, to the observed liquid-vapor transition in real gases, so that it is possible to claim that the van der Waals equation, in a sense, accounts for the existence of the liquid state. As the temperature is raised the amplitude of the "undulation" decreases and the volumes at the maximum and minimum come closer together. A temperature can be reached at which the maximum and minimum coalesce and the oscillation reduces to an inflection point at which the P-V isotherm has zero slope. This inflection point must correspond to the critical point of the gas, at which $P = P_c$, $V = V_c$, and $T = T_c$. The critical point is thus defined by the two conditions,

$$\text{zero slope of the isotherm} \quad \frac{dP}{dV} = 0 \tag{1-55}$$

$$\text{inflection point of the isotherm} \quad \frac{d^2P}{dV^2} = 0 \tag{1-56}$$

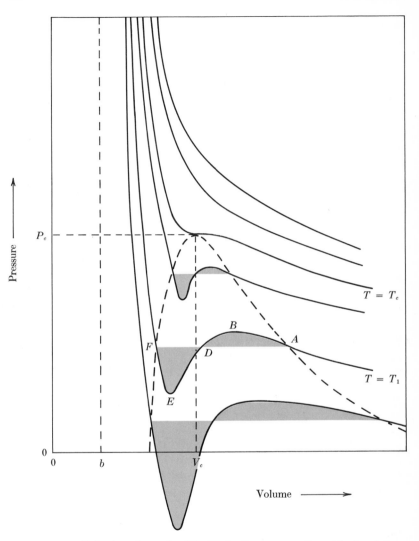

FIG. 1-7 *Behavior of van der Waals' isotherms near the critical point.*

EQUATIONS OF STATE OF GASES

When Eq. (1-54) is differentiated we find the following relations defining the critical point:

$$\frac{dP}{dV} = -\frac{RT_c}{(V_c - b)^2} + \frac{2a}{V_c^3} = 0 \quad \text{or} \quad \frac{RT_c}{(V_c - b)^2} = \frac{2a}{V_c^3} \quad (1\text{-}57)$$

$$\frac{d^2P}{dV^2} = \frac{2RT_c}{(V_c - b)^3} - \frac{6a}{V_c^4} = 0 \quad \text{or} \quad \frac{RT_c}{(V_c - b)^3} = \frac{3a}{V_c^4} \quad (1\text{-}58)$$

Dividing Eq. (1-58) by Eq. (1-57) we obtain the result

$$V_c - b = \frac{2}{3} V_c \quad \text{or} \quad V_c = 3b \quad (1\text{-}59)$$

and substituting this into Eq. (1-57), we find

$$\frac{RT_c}{(3b - b)^2} = \frac{2a}{27b^3} \quad \text{or} \quad T_c = \frac{8a}{27Rb} \quad (1\text{-}60)$$

Substituting both Eqs. (1-59) and (1-60) into Eq. (1-54) we find

$$P_c = \frac{RT_c}{V_c - b} - \frac{a}{V_c^2} = \frac{R \cdot 8a}{2b \cdot 27Rb} - \frac{a}{9b^2} = \frac{a}{27b^2} \quad (1\text{-}61)$$

Insofar as the van der Waals equation is an accurate representation of the true equation of state of a gas, it should therefore be possible to predict the critical constants, P_c, V_c, and T_c from the van der Waals constants a and b and the gas constant R. Or we might hope to be able to solve Eqs. (1-59), (1-60), and (1-61) to obtain expressions for the van der Waals constants a and b in terms of the critical constants; it is easily seen that this produces the result

$$b = \frac{V_c}{3} \quad a = 3P_c V_c^2 \quad R = \frac{8P_c V_c}{3T_c} \quad (1\text{-}62)$$

The values of the van der Waals constants a and b that one finds tabulated in handbooks are, in fact, usually determined in this way. The van der Waals equation is, however, rather unreliable quantitatively, especially in the vicinity of the critical point, and it would be very dangerous to use the values of a and b determined from Eq. (1-62) in order to obtain, say, the virial coefficients for use at pressures much below the critical pressure and at temperatures very different from the critical temperature. One indication of this unreliability is seen from Eq. (1-62), which gives for the compressibility factor at the critical point, according to the van der Waals equation,

$$\frac{P_c V_c}{RT_c} = \frac{3}{8} = 0.375 \tag{1-63}$$

Table 1-3 shows that this quantity in real gases usually has values of about 0.25 to 0.30, or only 70 to 80% as large as predicted by the van der Waals equation.

The student can show that the Berthelot equation of state leads to the results

$$V_c = 3b \qquad T_c = \left(\frac{8a}{27bR}\right)^{1/2} \qquad P_c = \left(\frac{aR}{216b^3}\right)^{1/2}$$

$$a = 3P_c V_c^2 T_c \qquad b = \frac{V_c}{3} \tag{1-64}$$

$$\frac{P_c V_c}{RT_c} = \frac{3}{8}$$

and the Dieterici equation gives (note that $e = 2.7182$)

$$V_c = 2b \qquad T_c = \frac{a}{4bR} \qquad P_c = \frac{a}{4e^2 b^2} = \frac{a}{29.6 b^2}$$

$$a = e^2 P_c V_c^2 \qquad b = \frac{V_c}{2} \tag{1-65}$$

$$\frac{P_c V_c}{RT_c} = \frac{2}{e^2} = 0.271$$

Thus the Dieterici equation gives a somewhat better value for the compressibility factor at the critical point than do the van der Waals and Berthelot equations. The student should recall, however, that the Berthelot equation gave a better representation of the temperature variation of the second virial coefficient; on the whole it is difficult to say which of the three equations gives the best overall representation of the true equation of state. Certainly none of them is perfectly satisfactory.

EQUATIONS OF STATE OF GASES

It is interesting to replace the constants a, b, and R in the van der Waals equation by the expressions given in Eq. (1-62),

$$P = \frac{8P_c V_c}{3T_c} T \frac{1}{(V - V_c/3)} - \frac{3P_c V_c^2}{V^2} \tag{1-66}$$

On rearrangement this gives

$$\left[\frac{P}{P_c} + 3\left(\frac{V_c}{V}\right)^2\right]\left[\frac{V}{V_c} - \frac{1}{3}\right] = \frac{8}{3}\frac{T}{T_c} \tag{1-67}$$

This equation assumes a remarkably simple form if one defines the so-called *reduced variables*,

$$P_r = \frac{P}{P_c} \qquad V_r = \frac{V}{V_c} \qquad T_r = \frac{T}{T_c} \tag{1-68}$$

Substitution of the reduced variables into Eq. (1-67) gives

$$\left(P_r + \frac{3}{V_r^2}\right)\left(V_r - \frac{1}{3}\right) = \frac{8}{3} T_r \tag{1-69}$$

an equation which does not contain explicitly any empirical constants. According to this equation, if we were to express the equation of state, $P = f(V, T)$ in terms of reduced variables, then the resulting equation of state,

$$P_r = F(V_r, T_r) \tag{1-70}$$

should be the same for all substances. Stated in another way, we can say that at a given reduced volume and reduced temperature, all gases and liquids must have the same reduced pressure. This principle is known as the *law of corresponding states*. In the way in which we have deduced it, the law appears to be a consequence of the van der Waals equation of state, but it actually has a more general theoretical basis than this. (The student can, for instance, show without great difficulty that the law would also be a consequence of either the Berthelot or Dieterici equations of state.)

A good test of the law of corresponding states can be made by plotting the compressibility factor, PV/nRT, against the reduced pressure for a number of gases at several reduced temperatures. This is shown in Fig. 1-8, which shows that the law is obeyed quite well—though not precisely—by substances as chemically distinct as water, nitrogen and carbon dioxide at temperatures above the critical temperature. Deviations from the law become a bit larger at temperatures below the critical temperature.

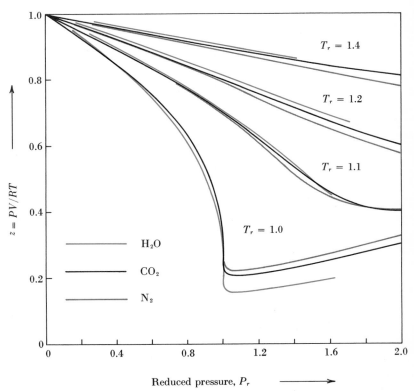

FIG. 1-8 *Validity of the law of corresponding states. The compressibility factors of nitrogen, carbon dioxide, and steam are plotted against the reduced pressure at several reduced temperatures near the critical point.* [Taken from G. N. Lewis and M. Randall, Thermodynamics, 2nd Ed. (revised by K. S. Pitzer and L. Brewer), McGraw-Hill, New York, 1961, p. 196.]

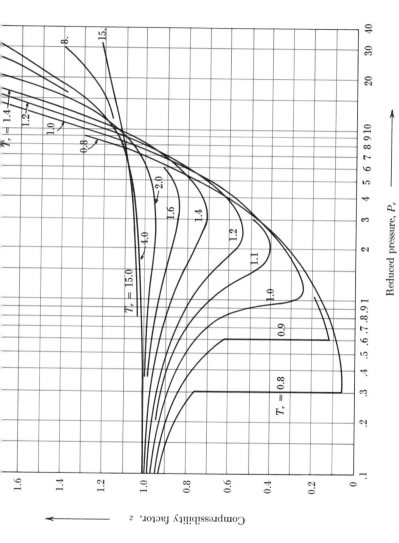

FIG. 1-9 *Dependence of the compressibility factor on reduced pressure at various temperatures.*

Reduced compressibility factor plots such as these provide a convenient and reasonably accurate means of estimating corrections to the ideal gas law and they are extensively used in practical work. A set of such plots is shown in Fig. 1-9.[13] As an example of the usefulness of such a plot, we may ask what is the actual mass of oxygen contained in a gas cylinder whose volume is 10 liters if the pressure of the gas is 100 atm and its temperature is 200°K. The critical pressure of oxygen is 49.7 atm and its critical temperature is 153°K. Thus $P_r = 100/49.7 = 2.006$, and $T_r = 200/153.4 = 1.30$. Reference to Fig. 1-9 shows that $PV/RT = 0.70$ under these conditions so that, taking $R = 0.082$ liter-atm/deg, $V = 0.70 \times 0.082 \times 200°/100$ atm $= 0.115$ liter/mole. The cylinder therefore contains $10/0.115 = 87$ moles $= 2.8$ kg of oxygen. (The ideal gas law would have predicted 2.0 kg.)

1-5 *Extensive and intensive properties*

The properties of the matter in a system can be classified into two groups depending on the manner in which they vary when the size of the system is changed. Let us suppose that we are given 100 g of water and that we measure some of its properties, such as volume, V, mass, m, density, d, vapor pressure, P, freezing point, T_m, heat evolved, Q_f, when the entire sample is frozen, and the volume change, ΔV_t, when the temperature is raised from 20 to 21°C. Next suppose that these same measurements are repeated under the same conditions except that our measurements are made on 200 g of water. It is evident that in these two sets of measurements the numerical values obtained for certain of the above-mentioned properties (namely, d, P, and T_m) will not differ, whereas the numerical values obtained for V, m, Q_f, and ΔV_t for the 200-g sample will be double the values obtained for the 100-g sample. Properties such as the density, vapor pressure and freezing point, which do not depend on the size of the sample of the substance studied, are called *intensive properties*. Properties such as volume, mass, heat evolution on freezing and volume change on heating by one degree, whose numerical values are proportional to the amount

---

[13] Constructed from the Hougen-Watson chart [O. A. Hougen and K. M. Watson, *Chemical Process Principles*, Wiley, New York, 1947, Part II] and the tables given by Pitzer *et al.* [*J. Am. Chem. Soc.*, 77, 3427(1955)]. Large scale plots are given in O. A. Hougen, K. M. Watson, and R. A. Ragatz, *Chemical Process Principles*, 2nd ed., Wiley, New York, Vol. II, 1960, pp. 573 ff.

of material in the sample under investigation, are called *extensive properties*. The ratio of two extensive properties is always an intensive property (for instance, $d = m/V$, the specific heat of fusion of ice $= Q_f/m$ and the volume coefficient of thermal expansion of water $= (dV/dt) \times (1/V)$). Extensive properties are attributes of particular systems (the water in the glass from which you are about to drink or the weight of carbon dioxide in the fire extinguisher in the organic chemistry laboratory) whereas intensive properties can be properties of substances in general (all water at 25°C and one atmosphere pressure has a density of 0.99707 g/ml and a volume coefficient of thermal expansion of 0.00026°C^{-1}).

Problems

1. The spectra of benzene (C_6H_6) liquid and vapor are to be compared at 30°C. A light beam from a spectrometer monochromator passes through a one centimeter layer of benzene (density 0.879 g/ml) and the light absorbed is measured. Then the same beam is passed through a layer of benzene vapor (vapor pressure 119 mm Hg at 30°C) of sufficient thickness to allow the light to interact with the same number of benzene molecules as are present in the one centimeter layer of the liquid. What must the thickness of the benzene vapor layer be?

2. A standard compressed gas cylinder has an interior length of 4 ft and an inside diameter of 7.5 in. When used to transport CO_2 it is filled to a pressure of 1800 lb/in.^{2} at room temperature (70°F). If CO_2 were an ideal gas, what weight of gas would the tank contain? What is the best estimate you can make of the actual weight of CO_2 in the tank? (You may use the critical constants given in Table 1-3 along with any Figures in this Chapter.)

3. Show that the second virial coefficient obtained from the Berthelot equation of state is $B = b - (a/RT^2)$ [cf. Eq. (1-48)].

4. Show that the second virial coefficient obtained from the Dieterici equation is $B = b - (a/RT)$ [cf. Eq. (1-49)].

5. Derive Eqs. (1-64) from the Berthelot equation of state.

6. Derive Eqs. (1-65) from the Dieterici equation of state.

7. Show that when reduced variables, $T^* = T/T_c$, $P^* = P/P_c$, and $V^* = V/V_c$ are used in the Berthelot equation of state, one obtains the reduced equation of state

$$\left(P^* + \frac{3}{T^*V^{*2}}\right)\left(V^* - \frac{1}{3}\right) = \frac{8T^*}{3}$$

8. Show that the Dieterici equation of state, when expressed in

reduced variables, has the form

$$P^* \exp\left(\frac{2}{V^*T^*}\right)\left(V^* - \frac{1}{2}\right) = \frac{e^2}{2} T^*$$

9. The *Boyle temperature* of a gas is the temperature at which the second virial coefficient vanishes. (a) Show that for a van der Waals gas, $T_{\text{Boyle}} = 9P_c V_c/R$, where P_c and V_c are the critical pressure and volume, respectively, and R is the gas constant. (b) The observed Boyle temperatures of some common gases are as follows: He, 35°K; H_2, 120°K; Ne, 120°K; N_2, 323°K; CO, 342°K; O_2, 423°K; A, 410°K, CO_2, 650°K. Compare these values with those predicted from the van der Waals equation using the values of P_c and V_c given in Table 1-3. (c) Show that from the van der Waals equation one can also predict that $T_{\text{Boyle}} = 27T_c/8$. Compare this prediction with the observed values.

10. Find the relationships between the Boyle temperature and the critical constants according to the Berthelot and Dieterici equations. Using the critical constants in Table 1-3, calculate the Boyle temperatures of the gases listed in Problem 9. Do these equations give better values for the Boyle temperatures than does the van der Waals equation?

11. The boiling point of benzene is 80.1°C and its density at its boiling point is 0.83 g/ml. Estimate the critical constants of benzene.

12. (a) The dirigible "Hindenberg" had a gas capacity of 7,060,000 cu ft. Compare the lifting powers at 1 atm and 70°F when filled with hydrogen and when filled with helium. (Remember the principle of Archimedes—that bouyancy is equal to the weight of the displaced fluid.) (b) The "Hindenberg" could ascend to a maximum altitude of 25,000 ft when filled with hydrogen. The air temperature at this altitude is 0°F and the pressure is $\frac{1}{3}$ atm; assume that the hydrogen temperature and pressure also have these values. Estimate the weight of the "Hindenberg" if it were filled with air at 1 atm and placed on a set of scales at sea level.

13. Instruments for studying cosmic rays are to be raised by balloon to an altitude of 100,000 ft. The instrument package weighs 50 kg and the balloon is to be made of plastic weighing 2 mg/cm². The temperature at 100,000 ft is -35°C and the atmospheric pressure is 0.01 atm. (a) Assuming that the balloon will be spherical when it reaches 100,000 ft, what is its diameter, and how much hydrogen must be used to fill it (in kilograms)? (b) If hydrogen comes in cylinders 4.5 ft long and 7.5 in. in diameter, at a pressure of 1,500 lb/in²(25°C), how many cylinders will be required to fill the balloon?

14. The change in pressure, dP, accompanying a change in altitude dh in a fluid in a gravitational field is governed by the equation $dP = -g\rho \, dh$, where g is the acceleration by gravity and ρ is the density of the fluid. Assuming that air is an ideal gas and that the temperature, T, of the

EQUATIONS OF STATE OF GASES

atmosphere does not vary with the altitude, show that the atmospheric pressure at altitude h is $P = P_0 \exp(-Mgh/RT)$, where P_0 is the pressure at sea level ($h = 0$), M is the mean molecular weight of the air, and R is the gas constant.

15. Two hundred years ago balloons were made which gained their bouyancy by being filled with heated air. Suppose that a deflated balloon weighs 100 lb and that when it is filled it forms a spherical shape 10 ft in diameter. Since the fabric of which the balloon is made is not very strong, the pressure inside and outside the bag must be the same. What is the minimum temperature that the air inside the bag must have in order for the balloon to rise 10,000 ft if the air outside the bag is at 60°F and if the pressure changes with altitude according to the equation given in Problem 14?

16. In quiet breathing the volume of air entering and leaving the lungs on each breath is about 500 ml and the breathing rate is about 20 breaths per minute. Body temperature is 37°C and the vapor pressure of water at this temperature is 33.7 mm Hg. Approximately how many glasses of water does one have to drink in one day in order to compensate for the loss of water from the lungs by breathing if the ambient temperature is 25°C and the ambient relative humidity is 50%? (The vapor pressure of water is 23.8 mm Hg at 25°C.)

17. Dry ice (solid CO_2) has been used as a mine explosive in the following way. A hole is drilled into the mine wall, filled with dry ice plus a small charge of gunpowder and then plugged. The gunpowder is lighted with a fuse, vaporizing the CO_2 and building up a high pressure within the hole, which breaks up the rock in the mine wall. (a) Calculate the pressure that will develop if 500 g of solid CO_2 is placed in a 1000 ml hole and warmed to 750°C, using the van der Waals equation and the critical constants of Table 1-3 to find a and b. (b) Calculate the pressure using the curves in Fig. 1-9.

18. From the following density vs. pressure data for CO_2 at 0°C find the molecular weight and the second virial coefficients B and B':

Pressure (atm)	Density (g/liter)
1.00000	1.97676
0.66667	1.31485
0.50000	0.98505
0.33333	0.65596

19. Show that if the equation of state of a gas is $[P + (a/T^m V^n)][V - b] = RT$, then the critical constants are

$$V_c = \frac{n+1}{n-1} b \qquad T_c^{m+1} = \frac{4na}{Rb^{n-1}} \frac{(n-1)^{n-1}}{(n+1)^{n+1}} \qquad P_c = \frac{(n-1)^2}{4n} \frac{RT_c}{b}$$

Supplementary references

J. R. Partington, *An Advanced Treatise on Physical Chemistry*, Wiley, New York, Vol. 1: *Fundamental Principles. The Properties of Gases*, 1949, pp. 546–729. An exhaustive and clearly written account of empirical equations of state of gases, thoroughly referenced.

J. A. Beattie, in *High Speed Aerodynamics and Jet Propulsion* (T. von Kármán, H. L. Dryden, and H. S. Taylor, eds.), Princeton University Press, Princeton, N. J., Vol. 1: *Thermodynamics and Physics of Matter* (F. D. Rossini, ed.), 1955, pp. 240–272.

G. N. Lewis and M. Randall, *Thermodynamics*, 2nd ed. (revised by K. S. Pitzer and L. Brewer), McGraw-Hill, New York, 1961, Chapter 16. Appendix I outlines a promising semi-empirical approach to equations of state.

Chapter 2

THE MOLECULAR EXPLANATION OF THE EQUATIONS OF STATE

SOME IMPORTANT properties of the equations of state of gases have been described in Chapter 1. We now consider how the equations, as well as other properties of gases, may be understood in terms of the molecular structure of a gas. This matter is dealt with in the so-called kinetic theory of gases. Although the basic assumptions of this theory are simple and plausible, the theory has been highly successful in accounting for the properties of gases, especially at low pressures. Consequently, the low pressure gaseous state is today more thoroughly understood in molecular terms than any other state of matter.

It is worth noting that until about the middle of the nineteenth century the generally accepted molecular explanation of Boyle's law was based on a notion proposed by Newton, who believed that gas molecules are fixed in position and that the pressure of a gas arises because the molecules repel each other. Scientists who followed Newton showed that Boyle's law would result if the repulsive force between a pair of molecules varies inversely with the distance between the molecules. These scientists realized, of course, that the molecules in a liquid or in a solid are much closer together than in a gas, so that they must for some reason attract one another, but it was felt that when solids or liquids are vaporized, the addition of the latent heat of vaporization ("caloric") in some way converted this attractive force into a repulsion.

2-1 Bernoulli's theory

The correct and quantitative basic picture of gas structure goes back to a chapter in the book, *Hydrodynamik*, by Daniel Bernoulli (1738), but even before this, Gassendi (1658) and Robert Hooke (ca. 1680) had the same idea in qualitative form. Interestingly enough, Bernoulli's theory was completely overlooked for over one hundred years. In 1821 an eccentric Englishman, John Herapath, independently hit upon a molecular model of a gas that is similar to Bernoulli's, but his reasons for proposing it now seem to be rather unconvincing. In 1845 John James Waterston, at that time a schoolteacher in Bombay, submitted a paper to the Royal Society in which many of the basic concepts of the kinetic theory were set forth. The paper was rejected as "nothing but nonsense, unfit even for reading before the Society." Soon thereafter the same ideas were developed in forceful fashion by Joule (1848), Rankin and Kronig (1856), and especially by Clausius (1857), Maxwell and Boltzmann. Bernoulli's contribution of 1738 was rediscovered in 1859 and the ideas originally proposed by him rapidly gained general acceptance.[1]

According to the Bernoulli theory, a gas is made up of a great many molecules moving chaotically through space in all directions. (In Waterston's words, a gas "may be likened to the familiar appearance of a swarm of gnats in a sunbeam.") The molecular size is small in comparison with the average distance between mole-

---

[1] An interesting—and chastening—sketch of the early history of the kinetic theory of gases is given by E. Mendoza, *Physics Today*, **14**, 36–39 (1961). Waterston's manuscript was found by Lord Rayleigh in 1892 in the archives of the Royal Society and was published in the *Philosophical Transactions of the Royal Society*. Lord Rayleigh's introductory comments on Waterston's paper make very interesting reading. Bernoulli's paper is translated in *The World of Mathematics*, J. R. Newman, ed., Vol. 2, Simon and Schuster, New York, 1956, p. 774.

cules, and the molecules exert no appreciable forces on one another. From these simple assumptions and Newton's laws of motion, Boyle's law is easily derived. Furthermore, important insight is gained into the molecular basis of temperature and heat. The assumptions concerning molecular sizes and intermolecular forces will be modified later in this chapter, where it is shown that these additional factors led van der Waals to his famous equation of state.

a. ELEMENTARY ACCOUNT OF BERNOULLI'S THEORY Let us now work out the physical consequences of Bernoulli's model of a gas. Assume that a box having the form of a rectangular parallelepiped of dimensions $a \times b \times c$ (see Fig. 2-1) contains N molecules, each of mass m. Consider one of these molecules (call it molecule 1). Let its velocity component parallel to the x-axis (which is parallel to the side of length a) be u_{x1}. Assuming that the components of motion in the other two directions do not influence the motion in the x-direction, molecule no. 1 will move back and forth, bouncing

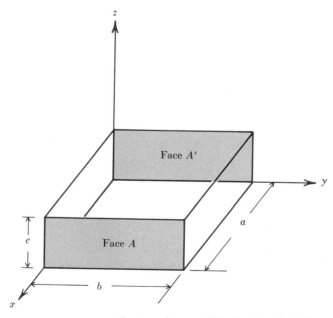

FIG. 2-1 *Rectangular parallelepiped used in the molecular explanation of Boyle's law.*

successively off the faces A and A' (shaded in Fig. 2-1). Assume for the moment that the collisions with the walls are perfect and that the molecule collides with no other molecules. (Both of these assumptions will be shown later to have no effect on our final result.) Between successive collisions with face A, the molecule travels a distance $2a$ in the x-direction, and the time interval between collisions is

$$\Delta t = \frac{2a}{u_{x1}} \tag{2-1}$$

provided that the molecule does not undergo collisions with other molecules in its travels back and forth in the container. On each collision the momentum component in the direction normal to face A changes from $+mu_{x1}$ to $-mu_{x1}$, a net momentum change

$$\Delta m u_{x1} = 2 m u_{x1} \tag{2-2}$$

Thus, taken over a long period of time, the momentum of the particle in the x-direction is changed at the average rate

$$\frac{\Delta m u_{x1}}{\Delta t} = \frac{2 m u_{x1}}{(2a/u_{x1})} = \frac{m u_{x1}^2}{a} \tag{2-3}$$

Newton's laws of motion assert that momentum can be changed only by exerting a force, and the average force exerted by molecule 1 on face A can be equated to the rate of change of momentum,

$$f_1 = \frac{\Delta m u_{x1}}{\Delta t} = \frac{m u_{x1}^2}{a} \tag{2-4}$$

The area of face A is bc, so that molecule 1 will exert on A an average pressure

$$P_1 = \frac{f_1}{bc} = \frac{m u_{x1}^2}{abc} \tag{2-5}$$

Since the product abc is the volume, V, of the box, we have

$$P_1 = \frac{m u_{x1}^2}{V} \quad \text{or} \quad P_1 V = m u_{x1}^2 \tag{2-6}$$

As long as u_{x1} does not change (and it will not if the particle undergoes perfect collisions with the wall), the quantity $m u_{x1}^2$ will be a constant, so we have

$$P_1 V = \text{const} \tag{2-7}$$

which has the form of Boyle's law.

EQUATIONS OF STATE OF GASES

Of course, if a single molecule were in a real box, one would expect to feel a succession of impulses on the wall, rather than a constant pressure. If the mass of the wall were M (assumed to be very much greater than the molecular mass, m) it would, on each collision, pick up an increment of velocity U in the x-direction, such that $MU = 2mu_{x1}$ (by the laws of momentum interchange on a perfect collision). After n collisions, the wall would move with a velocity of approximately nU, unless something were pushing against it with the constant force given by Eq. (2-4).

Each of the other molecules in the box will exert a similar pressure on face A, so we may write for the total pressure acting on face A

$$P = P_1 + P_2 + P_3 + \cdots + P_N$$
$$= \sum_{i=1}^{N} P_i = \sum \frac{mu_{xi}^2}{V} = \frac{m}{V} \sum u_{xi}^2 \qquad (2\text{-}8)$$

where P_i is the pressure exerted by molecule i, u_{xi} is the x-component of velocity of molecule i, and where the summation symbol has been introduced for the sake of brevity. We now define $\overline{u_x^2}$, the average value of u_x^2, by the equation[2]

$$\overline{u_x^2} = \frac{\Sigma u_{xi}^2}{N} \qquad (2\text{-}9)$$

[2] It is important to distinguish between $(\overline{u_x})^2$ and $\overline{u_x^2}$, where $\overline{u_x}$ is the average value of the x-component of velocity, defined by

$$\overline{u_x} = \frac{\Sigma u_{xi}}{N}$$

Since at any instant half of the molecules are moving toward face A and half are moving toward face A', half of the u_{xi} must be negative and half must be positive. Therefore $\overline{u_x}$ must be zero, corresponding to the fact that the gas as a whole is not moving. On the other hand each term in Σu_{xi}^2 is positive so that $\overline{u_x^2}$ is very different from zero. Thus

$$(\overline{u_x})^2 = 0 \neq \overline{u_x^2}.$$

The quantity $\overline{u_x}$ is called the "mean velocity component in the x-direction," and $\overline{u_x^2}$ is called the "mean square velocity component in the x-direction"; $(\overline{u_x})^2$ is called the "squared mean velocity component in the x-direction."

and we see that

$$P = \frac{mN\overline{u_x^2}}{V} \tag{2-10}$$

Now for each molecule

$$u_i^2 = u_{xi}^2 + u_{yi}^2 + u_{zi}^2 \tag{2-11}$$

where u_i is the speed of molecule i. Thus we can write

$$\begin{aligned} u_1^2 &= u_{x1}^2 + u_{y1}^2 + u_{z1}^2 \\ u_2^2 &= u_{x2}^2 + u_{y2}^2 + u_{z2}^2 \\ &\;\;\vdots \\ u_N^2 &= u_{xN}^2 + u_{yN}^2 + u_{zN}^2 \end{aligned}$$

If all of these equations are added and the result on both sides is divided by N, we obtain an expression for the mean square molecular speed,

$$\overline{u^2} = \frac{\Sigma \overline{u_i^2}}{N} = \frac{\Sigma u_{xi}^2}{N} + \frac{\Sigma u_{yi}^2}{N} + \frac{\Sigma u_{zi}^2}{N} = \overline{u_x^2} + \overline{u_y^2} + \overline{u_z^2} \tag{2-12}$$

At this point we introduce for the first time the assumption that the molecular motion is random, and that the three directions, x, y, and z, are entirely equivalent, so that

$$\overline{u_x^2} = \overline{u_y^2} = \overline{u_z^2} \tag{2-13}$$

This means that

$$\overline{u^2} = 3\overline{u_x^2} \tag{2-14a}$$

or

$$\overline{u_x^2} = \frac{\overline{u^2}}{3} \tag{2-14b}$$

Thus

$$PV = \frac{1}{3} Nm\overline{u^2} \tag{2-15}$$

EQUATIONS OF STATE OF GASES 55

If we compare this with the ideal gas law

$$PV = nRT \qquad (2\text{-}16)$$

we are led to the important identification

$$\frac{1}{3} Nm\overline{u^2} = nRT \qquad (2\text{-}17)$$

which relates the temperature to the molecular velocities in a manner consistent with the intuitive feeling that molecules move more rapidly when gases are heated. We shall regard Eq. (2-17) for the present as a reasonable hypothesis, which may be added to those made by Bernoulli. In the next chapter, however, we shall see that this equation provides the basis for a prediction about the specific heats of gases that is observed to be true. Thus Eq. (2-17) is really to be regarded as an interpretation of an experimental fact. It is also a key to the molecular interpretation of the concepts of temperature and heat.

If $\overline{u^2}$ is assumed to be independent of P and V when the temperature is held constant, then Eq. (2-15) gives Boyle's law. The Gay-Lussac and Avogadro laws follow provided we make the assumption given in Eq. (2-17).

The Bernoulli theory also readily explains the Dalton law of partial pressures. Suppose the gas consists of a mixture of N_1 molecules of mass m_1, N_2 molecules of mass m_2, ... Then if the mean square speed of molecules of mass m_n is $\overline{u_n^2}$, we can see from the same arguments that led to Eq. (2-15) that

$$PV = \frac{1}{3} \sum N_n m_n \overline{u_n^2} \qquad (2\text{-}18)$$

But if the container held only N_1 molecules of species 1, the pressure, or partial pressure, would be given by

$$P_1 = \frac{1}{3} \frac{N_1 m_1 \overline{u_1^2}}{V}$$

and similarly for the partial pressures of the other species

$$P_n = \frac{1}{3} \frac{N_n m_n \overline{u_n^2}}{V} \qquad (2\text{-}19)$$

Thus
$$PV = \sum P_n V$$
or
$$P = \sum P_n \qquad (2\text{-}20)$$

which is Dalton's law

In the course of the above derivation, two assumptions were made which are actually unnecessary:

(1) It was assumed that each collision of a molecule with the wall was perfect, so that during each collision the molecular momentum normal to the wall is merely reversed in sign. This cannot be the case because the walls of any real container are also made up of molecules which presumably undergo random thermal motion about the fixed points at which they are attached to other neighboring molecules. Collisions with such molecules can certainly be expected to result in changes in the magnitude of the velocity of the gas molecule normal to the wall. But if the temperature of the wall is the same as that of the gas, no thermal energy will be lost or gained on the average in collisions of the gas molecules with the wall. Therefore, on the average, the gas molecules must leave the wall with the same velocity that they had on approaching it. Thus Eq. (2-6) must be true when averaged over a large number of collisions.

(2) It was assumed that the molecules do not collide with each other. But if all molecules are identical, they will, in a sequence of collisions, tend to exchange the values of u_x that they had before the collisions. Therefore, although molecular collisions may change the number of times per second that a particular molecule strikes the wall, collisions cannot have any effect on the rate at which all molecules strike the wall. Thus this assumption is not necessary to the derivation.

EXERCISE Suppose that there were an attractive force between the gas molecules and the wall, so that as the gas molecules approach the wall they undergo an acceleration. Would this have any influence on the form of Boyle's law as derived by means of the Bernoulli theory? [*Hint:* What happens to the molecular speed when the molecules leave the wall after colliding with it?]

b. ALTERNATIVE TREATMENT OF BERNOULLI'S THEORY There is something artificial about the treatment just given, which assumes point-like molecules moving independently in a rectangular parallelepiped. It is therefore comforting to know that a somewhat more general justification

EQUATIONS OF STATE OF GASES

of Boyle's law can be given which does not depend on any assumption about the shape of the vessel containing the gas.

Consider a plane element of surface or wall with area A, adjacent to which is a gas consisting of molecules moving at random in the manner assumed by Bernoulli (Fig. 2-2). The element of area A can be of

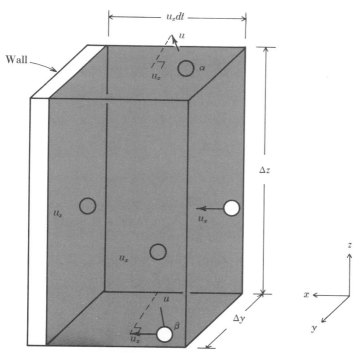

FIG. 2-2 *Collisions with an area $A = \Delta y\, \Delta z$ of molecules having x-velocity components between u_x and $u_x + du_x$. If the components of velocity in the y- and z-directions are disregarded, then all molecules lying within the shaded rectangular parallelepiped of dimensions Δy by Δz by $u_x\, dt$ will collide with the shaded surface in a time interval dt. This result is not changed if motions in the y- and z-directions are taken into account; we can expect that molecules, such as the molecule labeled α, which happen to leave the parallelepiped before collision with the wall because of the components of motion in directions normal to x, will on the average be exactly compensated by other molecules (such as the one labeled β) which enter the parallelepiped during time interval dt because of their motion in directions normal to x.*

macroscopic dimensions, provided that it is plane. Consider for the time being only those molecules in the gas whose component of velocity normal to the wall (let us call it the x-component) lies in the range u_x to $u_x + du_x$, where du_x is a small velocity increment. The number of molecules $d\nu(u_x)$ in one cubic centimeter having values of u_x lying in this range can be expected to be proportional to the width of the range, du_x, provided that this width is not too great, so we can write

$$d\nu(u_x) = \eta(u_x)\, du_x \tag{2-21}$$

where $\eta(u_x)$ is a proportionality factor whose numerical value will depend on u_x. (The function $\eta(u_x)$ is called a *distribution function*. Its functional form will be discussed in Chapter 4, but it is not necessary that we know it for the present discussion.) Since all molecules in one cubic centimeter must have some value of u_x lying between $-\infty$ and $+\infty$, it must be true that

$$\nu = \int_{-\infty}^{\infty} \eta(u_x)\, du_x \tag{2-22}$$

where ν is the total number of molecules in one cubic centimeter.

All molecules whose velocity is in the range u_x to $u_x + du_x$ and which lie within a distance $u_x\, dt$ of the wall will collide with the wall in a time interval dt (cf. Fig. 2-2) provided that u_x is directed toward the wall. If the positive x-direction is taken as pointing toward the wall, this means that collisions can occur only if $u_x > 0$. The number of molecules with this velocity which collide with the surface whose area is A in the time interval dt is equal to the number of molecules in a cylinder of base A and height $u_x\, dt$. Since $d\nu(u_x)$ is the number of molecules in one cubic centimeter having velocity components in the desired range, the number of molecules with velocity components in the range u_x to $u_x + du_x$ that collide with A is [3]

$$d\nu(u_x) \cdot A \cdot (u_x\, dt) = u_x A \eta(u_x)\, du_x\, dt \tag{2-23}$$

[3] An alert student might be concerned about the effects on this result of components of motion of gas molecules parallel to the surface (i.e., in the y- and z-directions); these motions could, in the time interval dt, bring molecules having the specified u_x into the cylinder through the sides. These molecules are capable of colliding with the area A of the wall in the time interval dt although they were not in the cylinder at the beginning of the interval. But the randomness of the motion of the molecules assures us that for every molecule entering the cylinder through the side in this way, another molecule must on the average leave the cylinder. Therefore Eq. (2-23) cannot be influenced by these motions parallel to the surface. Alternatively, we might have assumed that the linear dimensions of the cylinder base A are large compared with the height $u_x\, dt$, so that relatively few molecules with the specified u_x could enter and leave the cylinder through the sides, as compared with the number of molecules actually in the cylinder at the start of the interval dt.

EQUATIONS OF STATE OF GASES

Each of the molecules suffers a momentum change $2mu_x$ on collision, m being the molecular mass. Thus the change in momentum that occurs in time interval dt because of these collisions is

$$d(\text{momentum}) = 2mu_x \cdot u_x A \eta(u_x) \, du_x \, dt = 2mu_x^2 A \eta(u_x) \, du_x \, dt \quad (2\text{-}24)$$

and the contributions of these collisions to the pressure acting on the wall (that is, to the change of momentum divided by the area A of the element of the wall) is

$$dP = \frac{1}{A} \frac{d(\text{momentum})}{dt} = 2mu_x^2 \eta(u_x) \, du_x \quad (2\text{-}25)$$

The total pressure arising from collisions due to molecules moving with all possible positive velocity components u_x is (remember that molecules having negative values of u_x cannot collide with the wall, so they do not contribute to the pressure)

$$P = \int dP = \int_0^\infty 2mu_x^2 \eta(u_x) \, du_x = 2m \int_0^\infty u_x^2 \eta(u_x) \, dx \quad (2\text{-}26)$$

The mean square x-velocity component is given by the mean value principle (see Section 4-4 for a justification of this principle),

$$\overline{u_x^2} = \frac{\int_{-\infty}^{\infty} u_x^2 \eta(u_x) \, du_x}{\int_{-\infty}^{\infty} \eta(u_x) \, du_x} = \frac{1}{\nu} \int_{-\infty}^{\infty} u_x^2 \eta(u_x) \, du_x$$

$$= \frac{1}{\nu} \left[\int_{-\infty}^{0} u_x^2 \eta(u_x) \, du_x + \int_0^{\infty} u_x^2 \eta(u_x) \, du_x \right] \quad (2\text{-}27)$$

Because of the randomness of the motion of the gas molecules, the relative number of molecules moving with x-velocity component u_x must be equal to the relative number moving with x-velocity component of opposite sign, $-u_x$. Thus

$$\eta(u_x) = \eta(-u_x) \quad (2\text{-}28)$$

Furthermore

$$(u_x)^2 = (-u_x)^2 \quad (2\text{-}29)$$

so that

$$\int_{-\infty}^{0} u_x^2 \eta(u_x) \, du_x = \int_{\infty}^{0} (-u_x)^2 \eta(-u_x) \, d(-u_x)$$

$$= \int_0^{\infty} (-u_x)^2 \eta(-u_x) \, du_x$$

$$= \int_0^{\infty} u_x^2 \eta(u_x) \, du_x \quad (2\text{-}30)$$

which makes it possible to write Eq. (2-27) in the form

$$\overline{u_x^2} = \frac{2}{\nu} \int_0^\infty u_x^2 \eta(u_x) \, du_x \tag{2-31}$$

Substituting the integral in Eq. (2-26) from Eq. (2-31) we obtain

$$P = \nu m \overline{u_x^2} \tag{2-32}$$

If the gas contains a total of N molecules in a volume V, then

$$\nu = \frac{N}{V} \tag{2-33}$$

giving

$$P = \frac{Nm\overline{u_x^2}}{V} \tag{2-34}$$

which is identical with Eq. (2-10). The remaining arguments in Section 2-1a leading from Eq. (2-10) to Eq. (2-15) apply equally well to the present situation, so we may consider that we have derived Boyle's law for a container of arbitrary shape.

C. ESTIMATION OF AVERAGE MOLECULAR VELOCITIES; GRAHAM'S LAW OF EFFUSION If a container holds exactly one mole of gas and if the number of molecules in one mole is N_0, we can write from Eq. (2-17)

$$\frac{1}{3} N_0 m \overline{u^2} = RT \tag{2-35}$$

But $N_0 m$ is the mass of one mole of gas, which by definition is the molecular weight, M

$$N_0 m = M \tag{2-36}$$

Thus

$$\frac{1}{3} M \overline{u^2} = RT \tag{2-37}$$

or

$$\overline{u^2} = \frac{3RT}{M} \tag{2-38}$$

and we may calculate the mean square velocity of a gas directly from macroscopically observable quantities. The quantity $(\overline{u^2})^{1/2}$

is called the *root-mean-square velocity* (often denoted "rms velocity" or u_{rms}) [4]

$$u_{rms} = (\overline{u^2})^{1/2} = \sqrt{3RT/M} \tag{2-39}$$

Thus at a given temperature gases with the lowest molecular weights have the highest molecular velocities. For hydrogen, the lightest of all gases, we find at room temperature (approximately 300°K)

$$u_{rms} = [3 \times 8.314 \times 10^7 (\text{ergs/°K mole})$$
$$\times 300°K/2(\text{g/mole})]^{1/2}$$
$$= [37.40 \times 10^9 (\text{ergs/g})]^{1/2}$$
$$= 1.93 \times 10^5 \text{ cm/sec} = 1930 \text{ m/sec}$$

Note that if $(\overline{u^2})^{1/2}$ is to be given in the cgs units of cm/sec, then R and M must also be given in cgs units. The student should recall that 1 erg = 1 g (cm/sec)^2 so that $\sqrt{1 \text{ erg/g}} = 1$ cm/sec. For oxygen at room temperature the rms velocity is less than that of hydrogen by the factor $\sqrt{\frac{2}{32}} = \frac{1}{4}$, so that $u_{rms} = 483$ m/sec. These velocities are of the same order of magnitude as the velocities of sound in these gases (1332 m/sec for H_2, 333 m/sec for O_2) which suggests that sound propagation through a gas must be in some way associated with molecular motions and supports the validity of Eqs. (2-17) and (2-39). The molecular velocity of a gas varies relatively slowly with the temperature, a doubling of T producing an increase of only 41% in u_{rms}.

Equation (2-39) explains a phenomenon discovered by Graham in 1846. Let a gas be introduced into a container which has a small hole in its wall and let the space outside the container be evacuated (see Fig. 2-3). The hole must be sufficiently small that the molecule undergoes no collisions with other molecules in passing through it. (This will produce the result that the molecules will leave by *effusion* rather than by the flow of a stream or jet of gas.)

---

[4] In Chapter 4 it will be shown that the mean velocity, $\bar{u}$, is related to u_{rms} by a constant factor,

$$\bar{u} = \sqrt{8/3\pi}\, u_{rms}$$
$$= 0.921\, u_{rms}$$

so that u_{rms} gives a close approximation to the mean velocity.

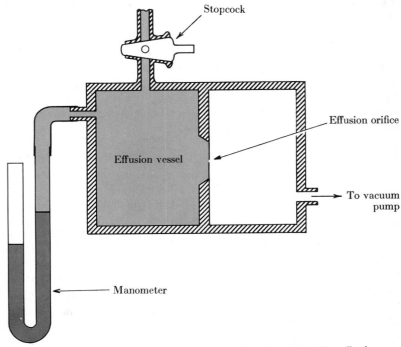

FIG. 2-3 *Effusion of gases. The gas is introduced into the effusion vessel through the stopcock, which is then closed. The rate at which the gas leaks or effuses through the orifice is measured by observing the rate at which the manometer pressure changes. After the rate has been measured the effusion vessel is evacuated and refilled through the stopcock with another gas.*

Then the rate at which gas leaves the container through the hole should be proportional to the average molecular velocity, which in turn would be expected to be inversely proportional to the square root of the molecular weight, because of Eq. (2-39). Therefore, if the effusion rates of two different gases through a given orifice at a given temperature and pressure are r_1 and r_2 (as measured by the rate of change of the manometer in Fig. 2-3) and if the molecular weights of the two gases are M_1 and M_2, we should expect to find

$$\frac{r_1}{r_2} = \left(\frac{M_2}{M_1}\right)^{1/2} \tag{2-40}$$

EQUATIONS OF STATE OF GASES 63

This relationship was observed by Graham at about the same time that the kinetic theory was beginning to be rediscovered and it provided strong evidence for its validity.

The effusion process may be used to separate the components of a gaseous mixture. If a mixture is placed in a container whose walls are porous, the constituent of lowest molecular weight should be able to leak through the walls most readily. Thus the gas remaining in the container after a period of time will be enriched in the heavier species, and the gas that has leaked through the wall will be enriched in the lighter species. This phenomenon was the basis of one of the processes developed in World War II at Oak Ridge for the manufacture of the uranium isotope 235 that was used in the atomic bombing of Hiroshima.

d. BOLTZMANN'S CONSTANT The number of molecules in a mole [N_0 in Eq. (2-35)] is called Avogadro's number and is known to have the value of 6.02252×10^{23} molecules per gram mole. If a sample of gas contains N molecules, then the number of moles is

$$n = \frac{N}{N_0} \tag{2-41}$$

Inserting this value of n in the ideal gas law we obtain

$$PV = \frac{N}{N_0} RT$$

It is convenient to define a new constant,

$$k = \frac{R}{N_0} \tag{2-42}$$

called the *Boltzmann constant*, or the *gas constant per molecule*. Then we can write the ideal gas law in the molecular form

$$PV = NkT \tag{2-43}$$

The Boltzmann constant has the numerical value, in cgs units, 1.38054×10^{-16} erg/°K molecule. We shall find that this constant appears very frequently in physical chemical equations based on molecular models.

2-2 The molecular explanation of deviations from the ideal gas law; van der Waals' theory

In deriving the ideal gas law from the Bernoulli theory, two assumptions were made which are not really correct. In the first place, molecules really do occupy space which, especially at higher densities, may not be negligible in comparison with the size of the container. Furthermore, molecules attract one another, and this influences the force exerted by molecules on the walls of the container. In 1873 the Dutch physical chemist, van der Waals, showed how these two factors could be introduced into the Bernoulli theory to give an equation of state. He then went on to show that this equation accounts semiquantitatively for the liquefaction of gases and the occurrence of the critical point. The arguments used by van der Waals in this stage of his theory have already been presented in Chapter 1.

a. EFFECT OF FINITE MOLECULAR VOLUME ON THE EQUATION OF STATE Van der Waals reasoned that, because of the finite size of the gas molecules, the space actually available for the motions of the molecules in a container is somewhat less than the total volume of the container. This has the effect that if the molecules are moving with a given mean velocity, there are more collisions of molecules with the walls in unit time than one would have found if the molecules were mathematical points. Of course, the finite size of the molecules may, especially at high densities, trap some of the molecules in the center of the container where they cannot collide at all with the wall, but this will be more than compensated by those molecules which are trapped near the walls, where they undergo repeated collisions with the walls. Van der Waals argued that for n moles of gas the volume V that appears in the Bernoulli derivation should be replaced by a corrected volume $(V - nb)$ where b is a constant called the *excluded volume* or *covolume* per mole of gas and n is the number of moles of gas present. The covolume b is related to the space occupied by the molecules themselves. Thus the finite molecular volume leads to the equation of state

$$P(V - nb) = nRT \qquad (2\text{-}44)$$

Van der Waals went on to show that if the gas molecules are rigid spheres, then the covolume b will be four times the space

EQUATIONS OF STATE OF GASES

actually occupied by the molecules in a mole of gas. That is,

$$b = \frac{2\pi}{3} N_0 d^3 \tag{2-45}$$

where d is the molecular diameter and N_0 is Avogadro's number. Since the volume v_0 of a sphere having a diameter d is

$$v_0 = \frac{\pi}{6} d^3 \tag{2-46}$$

we see that

$$b = 4N_0 v_0 \tag{2-47}$$

This result may be derived in the following way. Consider a particular molecule M in Fig. 2-4a. The center of molecule M will be unable to penetrate within the regions bounded by "exclusion spheres" of radius d drawn about all of the remaining $(N-1)$ molecules in the system. Each of these exclusion spheres has a volume eight times as great as the molecular volume, v_0, since the diameter of an exclusion sphere is double the molecular diameter (Fig. 2-4b). If the gas density is not too high, so that the mutual overlapping of the exclusion spheres of the molecules other than M can be neglected, then the center of M is excluded from a volume

$$B' = 8(N-1)v_0$$

where v_0 is given by Eq. (2-46). Since N is a very large number it is safe to write

$$B' = 8Nv_0$$

By the same argument the center of every other molecule in the gas is excluded from a similar volume. This means that the same excluded volume has been counted twice for each pair of molecules, so that the true excluded volume for all of the molecules in the gas must be

$$B = \frac{B'}{2} = 4Nv_0$$

Since

$$N = nN_0 \tag{2-48}$$

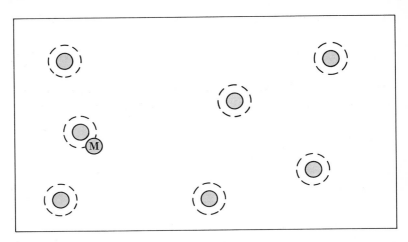

(a)

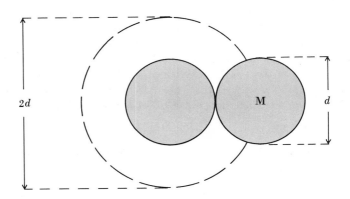

(b)

FIG. 2-4 (a) *The center of molecule M cannot be found within the spherical regions indicated by the dashed lines about the other molecules.* (b) *The volume about another molecule excluded to the center of molecule M is a sphere whose diameter is double the molecular diameter.* (c) *Boltzmann's evaluation of the van der Waals constant b.* (d) *Sharing of excluded volumes at high densities.*

EQUATIONS OF STATE OF GASES

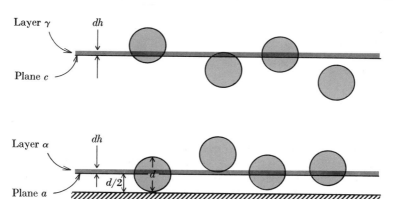

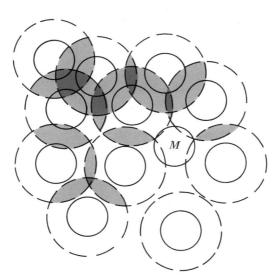

FIG. 2-4 *(Continued)*

and since the excluded volume B must be the same as nb in Eq. (2-44), we have

$$nb = 4nN_0v_0$$

or

$$b = 4N_0v_0 \tag{2-49}$$

which is Eq. (2-47).

Boltzmann has given an interesting alternative derivation of Eq. (2-44) which may seem more satisfactory to some students. (Still another derivation of this equation will be presented in Section 2-3b.) Consider the density of molecular centers in the layer of gas immediately adjacent to a macroscopic area A of the wall of the vessel containing the gas (A might be of the order of 1 cm^2 or so). It is obvious that no molecular centers can be found in the region between the wall and a plane a drawn parallel to the wall and a distance $d/2$ from the wall (Fig. 2-4c). The pressure on the wall is determined by the number density of molecular centers in the layer immediately adjacent to the plane a; it is the number density ν in this layer that appears in Eq. (2-32) in the calculation of the pressure in the Bernoulli theory. Boltzmann showed that for simple geometrical reasons the number density ν in this layer is not quite the same as the overall number density $\nu_0 = N/V$ of molecular centers in the body of the gas [cf. Eq. (2-33)].

In order to show this, Boltzmann compared the probability of finding the center of a particular molecule M in a layer α (Fig. 2-4c) of thickness dh and area A immediately adjacent to plane a with the probability of finding the center of molecule M in a layer γ of the same thickness and area adjacent to a plane c in the body of the gas far from a wall. The length dh is assumed to be small compared with the molecular diameter d. The space excluded to the center of molecule M in the entire volume V of the gas vessel is, from the discussion given above, $4\pi Nd^3/3$. The volumes of layers α and γ are both $A\,dh$. Within the layer γ adjacent to plane c, the excluded volume must be $(4\pi Nd^3/3)(A\,dh/V)$. Note that this excluded volume arises from molecules other than M whose centers may lie on either side of layer γ. Because $dh \ll d$ and the area A is very large compared with the molecular cross section, the centers of half of these molecules will lie above layer γ whereas the centers of the other half of the molecules will lie below layer γ. On the other hand, it is impossible to find any molecular centers on the side of layer α toward the wall. The entire excluded volume in layer α must, therefore, arise from molecules whose centers lie above this layer. Therefore the excluded volume in layer α must be exactly half of the excluded volume in the layer γ, or $(2\pi Nd^3/3)(A\,dh/V)$. The probability of finding the center

EQUATIONS OF STATE OF GASES

of molecule M in layer γ at a given instant is therefore proportional to

$$P_\gamma = \frac{1}{Adh - (4\pi Nd^3/3)(Adh/V)} \tag{2-50}$$

whereas the probability of finding M's center in layer α is proportional to

$$P_\alpha = \frac{1}{Adh - (2\pi Nd^3/3)(Adh/V)} \tag{2-51}$$

The ratio of the number density ν of molecular centers in layer α to the number density ν_0 of molecular centers in layer γ is obviously P_α/P_γ, so that

$$\frac{\nu}{\nu_0} = \frac{Adh - (2\pi Nd^3/3)(Adh/V)}{Adh - (4\pi Nd^3/3)(Adh/V)} \tag{2-52}$$

Writing $b = 2\pi N_0 d^3/3$ and $n = N/N_0$ [cf. Eqs. (2-45) and (2-48)] we obtain

$$\frac{\nu}{\nu_0} = \frac{1 - (nb/V)}{1 - (2nb/V)} \tag{2-53}$$

and using the expansion (valid for $x < 1$)

$$(1 - x)^{-1} = 1 + x + x^2 + x^3 + \cdots \tag{2-54}$$

we obtain

$$\frac{\nu}{\nu_0} = \frac{1}{[1 - (2nb/V)][1 + (bn/V) + (bn/V)^2 + \cdots]}$$

$$= \frac{1}{1 - (nb/V) - (nb/V)^2 - \cdots} \tag{2-55}$$

If $V \gg nb$ we may neglect powers of nb/V higher than the first and obtain

$$\nu = \frac{\nu_0}{1 - (nb/V)} \tag{2-56}$$

Since $\nu_0 = N/V$ we have for the number density of molecular centers next to the wall

$$\nu = \frac{N}{(V - nb)} \tag{2-57}$$

Substitution into Eq. (2-32) gives for the pressure of a gas composed of hard spheres

$$P = Nm\overline{u_x^2}/(V - nb) \tag{2-58}$$

where b is four times the space actually occupied by one mole of gas molecules.

It should be pointed out that both of these calculations of the excluded volume will be in error at high densities, where the probability is great that a given molecule M will come close to two or more other molecules which are already close to each other (Fig. 2-4d). Under these conditions part of the volume excluded to M (the shaded regions in Fig. 2-4d) is shared by two or more of the other molecules. It is evident that at high densities the volume excluded to the center of M is less than $8Nv_0$. Thus the covolume b is really not independent of the volume V of the system, but will tend to decrease as V decreases. This effect will be particularly important at densities equivalent to the liquid state, so the van der Waals equation cannot be expected to be correct for liquids unless it is suitably modified.[5]

Furthermore, real molecules are not hard spheres, and their "diameters" depend on the violence of the collisions that they undergo, violent collisions giving smaller diameters than gentle ones. Since collisions will be more violent at higher temperatures than at lower temperatures, it is evident that we must expect the "constant" b to vary with the temperature as well as with the density, becoming smaller as the temperature is increased. The "softness" of real molecules may also be expected to have an important influence on the equation of state at very high pressures, where the external pressure squeezes the molecules close together.

b. INTERMOLECULAR ATTRACTIONS The effect of intermolecular attractions on the equation of state was dealt with by van der Waals in the following way. At the instant that a molecule (such as molecule M in Fig. 2-5) undergoes a collision with the wall it will be "pulled back" by other molecules in its vicinity because of the attractive forces that are known to exist between molecules—the so-called *van der Waals forces*. In this way the impact of the

---

[5] It can be shown (J. O. Hirschfelder, C. F. Curtiss, and R. B. Bird, *Molecular Theory of Gases and Liquids*, Wiley, New York, 1954, pp. 4, 156) that this volume dependence of the covolume for a collection of rigid spheres of volume v_0 is given by

$$b = b_0 - \frac{0.375 n\, b_0^2}{V} + \frac{0.0369 n^2\, b_0^3}{V^2} + \cdots \qquad (2\text{-}59)$$

where $b_0 = 4 N_0 v_0$.

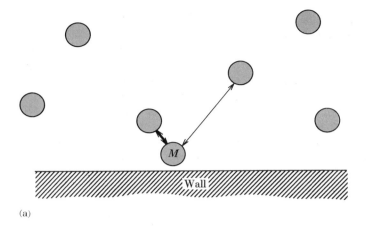

(a)

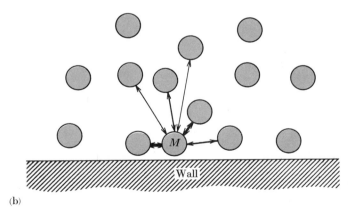

(b)

FIG. 2-5 *Reduction of collision impact with the wall because of intermolecular attractions. (a) Gas at low density. (b) Gas at high density. In the high density gas a particular molecule, M, suffers a smaller impact with the wall than in a low density gas because of the attractions of close neighbors, which reduce the force of the impact.*

collision of M with the wall is reduced because part of the force that causes the reversal of the momentum of molecule M at the wall arises in the gas itself. The average reduction in the impact of a single molecule may be expected to depend on the average number of other molecules in its immediate vicinity (compare Fig. 2-5a with Fig. 2-5b), which is in turn inversely proportional to V/n, the molar volume of the gas. The reduction in the total pressure arising from this effect will in turn be proportional to the number of molecules undergoing collision with unit area of the wall of the container at any moment, and this is also proportional to the gas density, or inversely proportional to the mole volume. Thus the reduction in the total pressure caused by molecular attractions is

$$P_{\text{red}} \propto \begin{pmatrix}\text{reduction of collision im-}\\ \text{pact of a single molecule}\end{pmatrix} \times \begin{pmatrix}\text{number of molecules}\\ \text{colliding with wall}\\ \text{in unit time}\end{pmatrix}$$

$$\propto \frac{n}{V} \times \frac{n}{V}$$

or

$$P_{\text{red}} = \frac{an^2}{V^2} \tag{2-60}$$

where a should be a constant independent of temperature and volume.

The arguments just outlined cannot be valid at very high densities and at very high pressures, where the molecules are squeezed together to such an extent that the pressure they exert on the walls comes largely from molecular deformation rather than from the momentum changes considered by Bernoulli. Furthermore, there is reason to expect a temperature dependence of the quantity a in Eq. (2-60) if the gas molecules have any tendency to form dimers, trimers, etc. It can be shown (see problem 6 at end of this chapter) that if a gas undergoes a dimerization reaction, $2A \rightleftharpoons A_2$, with the equilibrium constant

$$K = \frac{[A_2]}{[A]^2} \tag{2-61}$$

then, if n is the total number of molecules of A present, the pressure is

$$P = \frac{nRT}{V} - KRT\left(\frac{n}{V}\right)^2 + \text{terms of order } \left(\frac{n^3}{V}\right) \quad (2\text{-}62)$$

so that the reduction in pressure caused by dimerization is, to a first approximation,

$$P_{\text{red}} = \frac{KRTn^2}{V^2} \quad (2\text{-}63)$$

This reduction in pressure has the same dependence on (n/V) as does the reduction in pressure in Eq. (2-60), KRT playing the role of the quantity a in Eq. (2-60). The product KRT will be independent of temperature only if K is inversely proportional to T. This seems hardly likely; the temperature dependence of equilibrium constants is generally of the form

$$K = Ce^{-D/T} \quad (2\text{-}64)$$

where C and D are constants. Thus any contribution of dimerization to the negative term $a(n/V)^2$ in Eq. (2-60) will cause a to vary with temperature. Such a tendency to dimerization is to be expected in all gases since all gas molecules attract one another and therefore would be expected to exist (to some extent at least) in pairs, especially at low temperatures. Thus there is reason to expect that the van der Waals "constant" a must vary with both volume and temperature in all gases, at least over wide ranges of temperature and volume.

c. COMBINATION OF FINITE SIZE AND MOLECULAR ATTRACTION EFFECTS Van der Waals introduced the corrections we have just described into the Bernoulli theory in the following way. Let P_{ideal} be the pressure that the gas would exert if there were no intermolecular attractions and let V_{ideal} be the actual volume through which the centers of the gas molecules can move. Then the Bernoulli derivation shows that for n moles of gas,

$$P_{\text{ideal}} V_{\text{ideal}} = nRT \quad (2\text{-}65)$$

But van der Waals' considerations show that V_{ideal} is less than the actual volume, V, occupied by the gas by an amount nb,

$$V_{\text{ideal}} = V - nb \quad (2\text{-}66)$$

Furthermore P_{ideal} is greater than the observed pressure, P, by an amount an^2/V^2,

$$P_{\text{ideal}} = P + P_{\text{red}} = P + \frac{an^2}{V^2} \qquad (2\text{-}67)$$

Substituting Eqs. (2-66) and (2-67) into Eq. (2-65) we obtain the van der Waals equation,

$$\left(P + \frac{an^2}{V^2}\right)(V - nb) = nRT \qquad (2\text{-}68)$$

The success of this simple equation in accounting for the properties of real gases, and even to a limited extent of liquids, is truly amazing. The importance of the contribution made by van der Waals to the development of physical chemistry through this equation cannot be questioned. We have seen in Chapter 1 that it describes fairly well the observed deviations from ideality of real gases, and especially the temperature dependence of the second virial coefficient, the liquefaction of gases and the existence of the critical point. Nevertheless numerous assumptions were made in the derivation of the equation which limit its accuracy, and it cannot be relied upon to give a precise account of the behavior of real gases at high pressures and over wide ranges of temperature. For instance, the van der Waals equation predicts a lower limit, b, to the mole volume of a gas, at which volume the pressure will become infinite at all temperatures. Real substances possess no such minimum volume. This erroneous conclusion arises, of course, from the assumption that molecules are rigid objects with fixed boundaries—which they are not, even though they approximate to this in some degree.

Numerous efforts have been made to derive improved equations of state, especially at high pressures and for liquids. In the Berthelot equation, for instance, the constant a in van der Waals' equation is replaced by a/T. This gives a temperature dependence of the second virial coefficient that is in somewhat better agreement with experiment than the temperature dependence predicted by van der Waals' equation. In the Dieterici equation, the factor $P + an^2/V^2$ is replaced by $P \exp(a/VRT)$, which gives a value of the quantity $P_c V_c/RT_c$ that is closer to the observed value than the values obtained by either the Berthelot equation or the van der Waals equation. None of these equations are entirely adequate, however; they all have the difficulty that they predict a minimum volume, b, for the mole volume, at which the pressure becomes infinite.

The problem of obtaining a universally valid equation of state for gases is still unsolved. The problem of the equation of state of liquids is even farther from solution. These questions will be considered further in a subsequent volume.

2-3 *Statistical mechanical theory of the second virial coefficient*

a. THE POTENTIAL ENERGY FUNCTION FOR INTERMOLECULAR FORCES The qualitative ideas of fine molecular size and intermolecular attractions that van der Waals introduced may be very easily placed on a more quantitative basis. Let us imagine that two spherical molecules (say two argon atoms) are allowed to approach one another, and let us consider the forces that these molecules exert on each other. If the molecules are far apart we have every reason to believe that they exert no forces on each other at all. As they approach to a distance such that the molecular centers are a few molecular diameters apart, we may expect an attractive force to become noticeable, and this attractive force should increase as the intermolecular distance decreases. At some separation, however, the electronic clouds of the molecules will begin to overlap appreciably and, especially if they are inert atoms such as argon, the molecules will begin to resist any further decrease in their separation. The attractive force will be overwhelmed by this repulsive force arising from electron overlap, and the net force acting between the molecules will become repulsive. If one tries to push the molecules still closer together, the repulsion will greatly increase, presumably going toward infinity as the intermolecular distance is reduced to zero.

This behavior is sketched in Fig. 2-6. The physical origin of the attractive forces between molecules such as argon was explained by London in 1933. He considered a pair of molecules separated by a sufficient distance that their electron clouds do not overlap appreciably. He showed that there is a tendency for the electrons in one molecule to correlate their motions with those of the electrons in the other molecule in such a way that electrical fields are produced which cause the molecules to attract one another. This attractive force is called a *dispersion force* or a *London force*. London proved that this attractive force—which we shall denote by f_L—varies inversely with the seventh power of the intermolecular distance, r,

$$f_L = -\frac{K}{r^7} \tag{2-69}$$

where K is a positive constant whose value depends on the nature

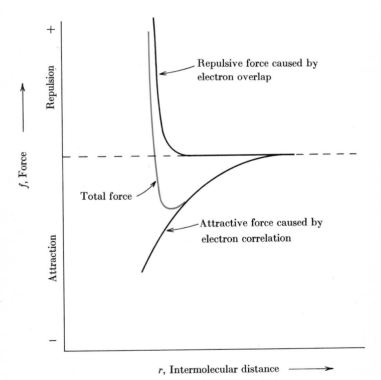

FIG. 2-6 *General character of the variation of the intermolecular force, f, with intermolecular distance, r, for a pair of spherical molecules.*

of the two molecules. The negative sign given to f_L indicates that the force is attractive.[6]

Less is known about the dependence on the intermolecular distance of the repulsive forces, but Lennard-Jones found that a function of the form

$$f_R = \frac{L}{r^m} \tag{2-71}$$

is both a satisfactory representation of the true repulsive force and a convenient force law for algebraic manipulation. In Eq. (2-71) L and m are empirical constants. In order for the repulsive forces to predominate over the attractive forces at small distances, the exponent m in Eq. (2-71) must have a value greater than 7, the value of the exponent in London's formula [Eq. (2-69)]. Lennard-Jones showed that a value of m of the order of 13 to 15 seems to be most nearly consistent with the experimental observations. For reasons given below, it is convenient to choose $m = 13$ in Eq. (2-71).

Thus, the general dependence of the intermolecular forces between spherical molecules is of the form

$$f = \frac{L}{r^{13}} - \frac{K}{r^7} \tag{2-72}$$

---

[6] London showed that the numerical value of K for a pair of molecules A and B can be estimated to a good approximation by the relation

$$K = \frac{9 I_A I_B \alpha_A \alpha_B}{(I_A + I_B)} \tag{2-70}$$

where α_A and α_B are the polarizabilities of the two molecules and I_A and I_B are their ionization energies. Molecular polarizabilities are a measure of the ease with which the electronic orbits in the molecule may be distorted by an external electric field. They can be determined from measurements of refractive indices. The ionization energy is the energy required to remove an electron from a molecule. It can be determined from the analysis of molecular spectra or from electron impact studies. Ionization energies of the common gas molecules almost invariably fall in the range 10^{-11} to 4×10^{-11} ergs, but the polarizabilities of different molecules may differ by a much larger factor. (The polarizability of helium is 0.67×10^{-24} cm^3, whereas that of iodine is approximately 11×10^{-24} cm^3; the ionization energies of these two gases are 3.9×10^{-11} and 1.55×10^{-11} ergs, respectively.) Thus the differences between the K values for different gases are predominantly determined by their polarizabilities, polarizable molecules attracting one another more strongly than nonpolarizable ones.

It is convenient to express this force law in terms of the potential energy, $V(r)$, required to bring two molecules from infinite separation to a distance r. The change in energy, dV, required to change the interatomic distance by dr is given by

$$dV = -f\, dr \qquad (2\text{-}73)$$

[The choice of sign on the right-hand side of this equation is dictated by our definition of the attractive force as a negative quantity, and the repulsive force as a positive quantity, in Eqs. (2-69) and (2-71). If two molecules attract (negative f), and if their separation is increased by dr—a positive quantity—then work is done *on* the molecule so the potential energy increases (positive dV).] Thus

$$\begin{aligned} V(r) &= -\int_\infty^r \left(\frac{L}{r^{13}} - \frac{K}{r^7}\right) dr \\ &= \frac{L}{12 r^{12}} - \frac{K}{6 r^6} \end{aligned} \qquad (2\text{-}74)$$

which can be written in the form

$$V(r) = \frac{a}{r^{12}} - \frac{b}{r^6} \qquad (2\text{-}75)$$

where a and b can be regarded as empirical parameters characteristic of any given molecular pair. The potential given in Eq. (2-75) is called the *Lennard-Jones 12-6 potential*. Note that $V(r)$ has been defined in such a way that $V(\infty) = 0$—which is a convenient, though arbitrary, choice of the zero of energy.

It is convenient to write the Lennard-Jones 12-6 potential in a slightly different form by replacing the two parameters a and b as follows. First we may define the interatomic distance σ at which the repulsive potential, a/r^{12}, is equal to the attractive, London potential, b/r^6. This distance is defined by the equation

$$V(\sigma) = \frac{a}{\sigma^{12}} - \frac{b}{\sigma^6} = 0 \qquad (2\text{-}76)$$

giving

$$a = b\sigma^6 \qquad (2\text{-}77)$$

EQUATIONS OF STATE OF GASES

On replacing a in Eq. (2-75) by $b\sigma^6$ we obtain

$$V(r) = b\sigma^{-6}\left[\left(\frac{\sigma}{r}\right)^{12} - \left(\frac{\sigma}{r}\right)^6\right] \tag{2-78}$$

Let us now determine the value of r at which the potential energy passes through its minimum value (this is, of course, also the distance at which the attractive and repulsive forces just balance one another). We find

$$\begin{aligned}dV/dr &= b\sigma^{-6}\left[-12\left(\frac{\sigma^{12}}{r^{13}}\right) + 6\left(\frac{\sigma^6}{r^7}\right)\right] \\ &= 6\left(\frac{b}{r^7}\right)\left[2\left(\frac{\sigma}{r}\right)^6 - 1\right]\end{aligned} \tag{2-79}$$

Setting $r = r_{\min}$ when $dV/dr = 0$ we obtain

$$r_{\min} = 2^{1/6}\sigma \tag{2-80}$$

The magnitude of $V(r)$ at $r = r_{\min}$ will be denoted by $-\epsilon$

$$V(r_{\min}) = -\epsilon = b\sigma^{-6}[2^{-2} - 2^{-1}] = -\frac{1}{4}b\sigma^{-6} \tag{2-81}$$

or

$$b\sigma^{-6} = 4\epsilon \tag{2-82}$$

so that we may write the Lennard-Jones 12-6 potential in the particularly simple form

$$V(r) = 4\epsilon\left[\left(\frac{\sigma}{r}\right)^{12} - \left(\frac{\sigma}{r}\right)^6\right] \tag{2-83}$$

The significance of the parameters ϵ and σ is evident from Fig. 2-7.

EXERCISE Given the Lennard-Jones $n - m$ potential

$$V(r) = \frac{a}{r^m} - \frac{b}{r^n}$$

show that if σ is the value of r at which the attractive and repulsive terms are equal, and if $-\epsilon$ is the minimum value of $V(r)$, then we can write

$$V(r) = A\epsilon\left[\left(\frac{\sigma}{r}\right)^m - \left(\frac{\sigma}{r}\right)^n\right] \tag{2-84}$$

where the distance, $r_{\min}$, at which $V(r)$ attains its minimum value is given by

$$r_{\min} = \left(\frac{m}{n}\right)^{1/(m-n)} \sigma \tag{2-85}$$

and where A is a constant given by

$$A = \frac{m}{m-n}\left(\frac{m}{n}\right)^{n/(m-n)} \tag{2-86}$$

If two nonspherical molecules, such as benzene, interact, the potential energy will depend on the relative spatial orientations of the molecules as well as on their distance apart. The potential function will thus depend on several variables in addition to the intermolecular distance and will assume a more complex form than

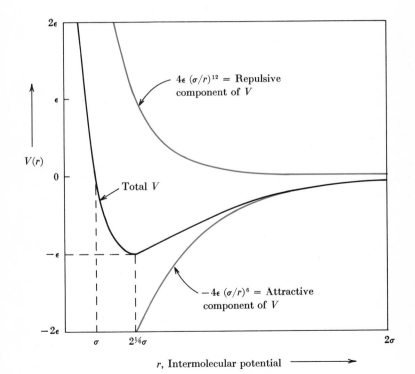

FIG. 2-7 *The Lennard-Jones 6–12 potential.*

EQUATIONS OF STATE OF GASES

that of the Lennard-Jones potential. For polar substances, such as water, which contain electric dipoles, new types of interaction occur in addition to the London forces and the repulsive forces described above. These forces add still more terms to the intermolecular potential.

b. THE RESULTS OF THE STATISTICAL MECHANICAL THEORY OF THE SECOND VIRIAL COEFFICIENT If the virial expansion is written in the form

$$\frac{PV}{nRT} = 1 + B(T)\frac{n}{V} + \cdots \qquad (2\text{-}87)$$

then it is possible to show by means of statistical mechanics that for spherical molecules the second virial coefficient, B, is related to the intermolecular potential, $V(r)$, just discussed, by the equation

$$B(T) = 2\pi N_0 \int_0^\infty (1 - e^{-V(r)/kT}) r^2 \, dr \qquad (2\text{-}88)$$

where N_0 is Avogadro's number and k is Boltzmann's constant. (This relationship will be derived in Volume 2.) This equation provides us with a means of gaining information about the potential functions of pairs of simple gas molecules from experimental measurements of the temperature variation of second virial coefficients. Much of our present knowledge of intermolecular forces has been obtained in this way.

It is instructive to apply Eq. (2-88) to the calculation of the second virial coefficients for various kinds of intermolecular potentials. First let us consider a gas in which the molecules are rigid spheres of diameter σ having no attractive forces for each other. The potential energy function appropriate to this kind of gas is

$$V(r) = \begin{cases} 0 & \text{if } r > \sigma \\ \infty & \text{if } r < \sigma \end{cases} \qquad (2\text{-}89)$$

Figure 2-8a shows the general shape of this potential and it also shows the dependence of the integrand of Eq. (2-88) on r. The second virial coefficient arising from this potential is given by

$$B(T) = 2\pi N_0 \left[\int_0^\sigma \left[1 - \exp\left(-\frac{\infty}{kT}\right)\right] r^2 \, dr \right.$$
$$\left. + \int_\sigma^\infty \left[1 - \exp\left(-\frac{0}{kT}\right)\right] r^2 \, dr \right] \qquad (2\text{-}90)$$

	(a) Hard sphere molecules	(b) Hard sphere plus attractions	(c) Lennard-Jones molecules
Intermolecular potential energy $V(r)$:	$V = \infty$ for $r < \sigma$		
$e^{-V(r)/kT}$:	Same at all temperatures	$T_1 > T_2 > T_3$	$T_1 > T_2 > T_3$

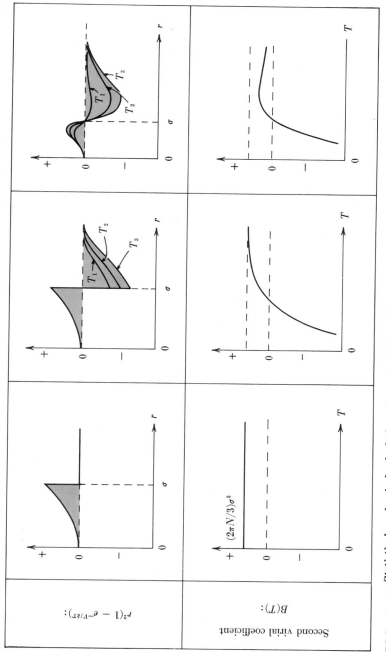

FIG. 2-8 *Statistical mechanical calculation of second virial coefficients with various simple models of the intermolecular potential.*

Since exp $(-\infty/kT) = 0$ at all temperatures and since exp $(-0/kT) = 1$ at all temperatures we have

$$B(T) = 2\pi N_0 \left[\int_0^\sigma r^2\, dr + \int_\sigma^\infty 0 \cdot r^2\, dr \right]$$
$$= \frac{2}{3} \pi N_0 \sigma^3 \tag{2-91}$$

Thus for gases composed of rigid spherical molecules the second virial coefficient is positive and independent of the temperature. The result obtained in Eq. (2-91) is precisely the value of the second virial coefficient obtained from van der Waals' equation of state if the contribution of the term (a/V^2), arising from intermolecular attractions, is neglected [cf. Eqs. (2-49) and (1-46), recalling that the volume per molecule is $v_0 = \pi\sigma^3/6$].

Let us next suppose that the gas consists of rigid spheres which attract one another if their centers are separated by a distance greater than a molecular diameter. If the attractive potential is inversely proportional to some power of the separation, we have the potential

$$V(r) = \begin{cases} \dfrac{-A}{r^n} & \text{if } r > \sigma \\ \infty & \text{if } r < \sigma \end{cases} \tag{2-92}$$

This potential, and the r-dependence of the integrand of Eq. (2-88) arising from it, are shown in Fig. 2-8b. It is clear that the contribution to the second virial coefficient by the integral over values of r between zero and σ is the same as that just found for the rigid nonattracting sphere case. On the other hand for r greater than σ the potential function $V(r)$ is negative. Thus the quantity $(-V/kT)$ is positive so that $\exp(-V/kT)$ is greater than unity and $[1 - \exp(-V/kT)]$ is a negative quantity. Thus there is a negative contribution to the second virial coefficient from the integral in Eq. (2-88) over values of r greater than σ. Furthermore, this portion of the integral is a larger negative number, the lower the temperature, since $\exp(-V/kT)$ increases as T decreases. Evidently the attractive forces tend to give negative contributions to the second virial coefficient and these contributions can be expected to predominate at low temperatures. The

EQUATIONS OF STATE OF GASES

expression for the second virial coefficient in this case is

$$B(T) = 2\pi N_0 \left[\int_0^\sigma r^2\, dr + \int_\sigma^\infty \left[1 - \exp\left(\frac{A}{r^n kT}\right)\right] r^2\, dr \right]$$
$$= \frac{2}{3}\pi N_0 \sigma^3 + 2\pi N_0 \int_\sigma^\infty \left[1 - \exp\left(\frac{A}{r^n kT}\right)\right] r^2\, dr \quad (2\text{-}93)$$

Unfortunately, the integral in the second term on the right in Eq. (2-93) cannot be evaluated in simple analytical form. An approximation is, however, possible at high temperatures, provided that $A/r^n kT \ll 1$ for $r > \sigma$. If this condition is satisfied, we can expand the exponential using the relation

$$e^x = 1 + x + \frac{1}{2}x^2 + \cdots \quad (2\text{-}94)$$

Disregarding all powers of the exponent beyond the linear term, we obtain

$$\exp\left(\frac{A}{r^n kT}\right) = 1 + \frac{A}{r^n kT} \quad (2\text{-}95)$$

$$\left. \begin{array}{l} \int_\sigma^\infty \left[1 - \exp\left(\frac{A}{r^n kT}\right)\right] r^2\, dr = -\frac{A}{kT}\int_\sigma^\infty r^{2-n}\, dr \\ \qquad = \frac{A}{(n-3)kT}\left[\frac{1}{r^{n-3}}\right]_\sigma^\infty \end{array} \right\} \quad (2\text{-}96)$$

This integral converges only if $n > 3$, giving the result

$$B(T) = \frac{2}{3}\pi N_0 \sigma^3 \left[1 - \frac{3A}{(n-3)\sigma^n kT}\right] \quad (2\text{-}97)$$

If we write $A/\sigma^n = \epsilon$, which is the value of $V(r)$ when the molecules are in contact, we obtain

$$B(T) = \frac{2}{3}\pi N_0 \sigma^3 \left[1 - \frac{3}{n-3}\frac{\epsilon}{kT}\right] \quad (2\text{-}98)$$

This expression for the second virial coefficient is of the same form as that obtained from the van der Waals theory,

$$B(T) = b - \frac{a}{RT} = b\left(1 - \frac{a}{bRT}\right) \qquad (2\text{-}99)$$

where

$$\left.\begin{aligned} b &= \frac{2}{3}\pi N_0 \sigma^3 \\ \frac{a}{b} &= \frac{3}{n-3}\frac{R}{k}\epsilon \end{aligned}\right\} \qquad (2\text{-}100)$$

If we assume that the attractive potential arises from the London forces, we may take $n = 6$, and recalling that $R = N_0 k$ we obtain

$$\frac{a}{b} = N_0 \epsilon \qquad (2\text{-}101)$$

Thus according to this approximation to the law of force between gas molecules we obtain simple expressions relating the parameters of the potential function to the van der Waals constants. The general character of the $B(T)$ vs. T plot is shown in Fig. 2-8b. It closely resembles the observed behavior except that B approaches a constant value at high temperatures whereas in real gases B passes through a weak maximum.

Lennard-Jones suggested that the more realistic potential, Eq. (2-84), be assumed and that the parameters ϵ, σ, m, and n be chosen to give the best possible fit to the observed temperature variation of B. In this way he was able to show that the most probable value of n, the exponent of $(1/r)$ in the attractive potential, is close to 6 and the value of m had to be of the order of 10 to 15. More recently, Hirschfelder and his collaborators have assumed $m = 12$, $n = 6$ and have obtained the values of ϵ and σ that give the best fit to the observed second virial coefficient versus temperature plots of numerous simple gases. Fig. 2-9 shows the nature of the agreement with the experimental observations obtained by Hirschfelder et al. This figure was constructed by assigning values of σ and ϵ to each of a number of gases, and plotting $B(T)/(\tfrac{2}{3}\pi N_0 \sigma^3)$ against $(k/\epsilon)T$. We may write

$$B(T) = 2\pi N_0 \int_0^\infty \left\{1 - \exp\left[-\frac{4\epsilon}{kT}\left(\left(\frac{\sigma}{r}\right)^{12} - \left(\frac{\sigma}{r}\right)^6\right)\right]\right\} r^2\, dr \qquad (2\text{-}102)$$

EQUATIONS OF STATE OF GASES

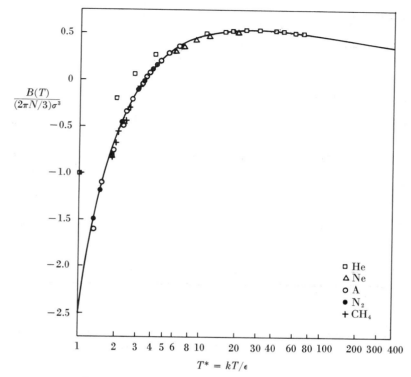

FIG. 2-9 *Test of the Lennard-Jones 6–12 potential as a basis for second virial coefficients of simple gases.*

Let us replace r in this integral by a new variable, $x = r/\sigma$. Furthermore, we may note that the quantity ϵ/k has the dimensions of temperature, so that we may define a dimensionless reduced temperature, $T^* = (k/\epsilon)T$. We then obtain from Eq. (2-102)

$$\frac{B(T)}{\frac{2}{3}\pi N_0 \sigma^3} = 3 \int_0^\infty \left\{1 - \exp\left[-\frac{4}{T^*}(x^{-12} - x^{-6})\right]\right\} x^2\, dx \tag{2-103}$$

The value of the integral here should depend only on the numerical value of the reduced temperature, T^*, so that two gases at the same reduced temperature must have the same values of B/σ^3.

TABLE 2-1 VALUES OF ϵ AND σ IN THE LENNARD-JONES 6-12 POTENTIAL DERIVED FROM SECOND VIRIAL COEFFICIENTS AND VISCOSITIES[a]

Gas	Boiling point, T_b (°K)	From second virial coef.			From viscosity	
		ϵ/k (°K)	σ (Å)	ϵ/kT_b	ϵ/k (°K)	σ (Å)
Ne	26.3	35.6	2.75	1.35	35.7	2.79
A	87.5	119.8	3.40	1.37	124	3.42
Kr	118.3	171	3.60	1.45	190	3.61
Xe	166.1	221	4.10	1.27	229	4.06
N_2	77.4	95.05	3.70	1.23	91.5	3.68
O_2	90.2	117.5	3.58	1.30	113	3.43
CO_2	194.7 (subl.)	205	4.07	1.05	190	4.00
CH_4	111.7	148.2	3.82	1.27	137	3.88

[a] From J. O. Hirschfelder, C. F. Curtiss, and R. B. Bird, *Molecular Theory of Gases and Liquids*, Wiley, New York, 1954, Table 1-A, pp. 1110–1112. Where more than one set of values is given in this table, the one based on the most recent measurements was selected. The procedure for deriving values of ϵ and σ from viscosity data will be discussed in Chapter 5.

Figure 2-9 shows that values of σ and ϵ can be chosen in such a way that this is indeed the case for a large number of simple gases.[7] Values of ϵ and σ that give good fits to the observed second virial coefficient variation with temperature are given in Table 2-1. It should be noted that the quantity ϵ/k is the temperature at which $\epsilon = kT$—that is, the temperature at which the mean thermal energy becomes of the same order of magnitude as the maximum potential energy of attraction between pairs of molecules. Table 2-1 shows that the temperature ϵ/k is approximately proportional to the boiling point of the gas at 1 atm, the constant of proportionality being about 1.3.

[7] Note that the points for helium in Fig. 2-9 deviate from the behavior of other gases at low temperatures. This is caused by quantum effects which are unimportant for other gases.

Problems

1. Two vessels, A and B, whose walls are maintained at temperatures T_A and T_B, respectively, are connected by a fine hole through which gas molecules can effuse back and forth from one vessel to the other. Prove that $P_A/P_B = (T_A/T_B)^{1/2}$ where P_A and P_B are the pressures in the two vessels when a steady state has been achieved.
2. An effusion vessel contains a mixture consisting of 90% hydrogen and 10% nitrogen. What is the composition of the gas that escapes through the effusion orifice?
3. Some oxygen is placed in a cubical box 10 cm on a side at a temperature of 25°C and at a pressure so low that the molecules collide with the walls far more often than they collide with each other. How many times per second on the average does a given oxygen molecule collide with a given wall?
4. Two gases, a and b, consist of rigid, spherical molecules, all of the same diameter. A volume V is filled with n_a moles of a and n_b moles of b. The molecules exert no attractive forces on each other. Would you expect the second virial coefficient, B, in the virial equation

$$\frac{PV}{(n_a + n_b)RT} = 1 + B\left(\frac{n_a + n_b}{V}\right) + \cdots$$

to vary with the composition of the gas?

5. Will the gases described in Problem 4 obey Dalton's law?
6. A gas undergoes a dimerization reaction, $2A \rightleftharpoons A_2$, with an equilibrium constant $K = [A_2]/[A]^2$. Assuming that excluded volume and intermolecular attraction effects (other than those leading to the dimerization) are negligible, show that [cf. Eq. (2-62)]

$$\frac{PV}{nRT} = 1 - K\left(\frac{n}{V}\right) + \text{terms of order } \left(\frac{n}{V}\right)^2$$

where n is the total number of moles of A added. [*Hint:* Note that $n/V = [A] + 2[A_2]$ and $P = ([A] + [A_2])RT$. Also, for small x, $(1+x)^{1/2} = 1 + \frac{1}{2}x - \frac{1}{8}x^2 + \cdots$.]

7. Find the third virial coefficient of the gas described in Problem 6.
8. It is found that the integral in Eq. (2-103) vanishes when $T^* = 3.42$. Use this fact to estimate ϵ/k from the Boyle temperatures given in Problem 9 of Chapter 1 and compare your estimates with the values given in Table 2-1.

Supplementary references

R. D. Present, *Kinetic Theory of Gases*, McGraw-Hill, New York, 1958, Chapter 16. An excellent recent text.

J. O. Hirschfelder, C. F. Curtiss, and R. B. Bird, *Molecular Theory of Gases and Liquids*, Wiley, New York, 1954, Chapters 3-6. A definitive discussion of the theory of the equation of state and its application to experiment. Chapters 12-14 contain an extensive discussion of intermolecular forces.

L. Boltzmann, *Lectures on Gas Theory* (translated by S. G. Brush), University of California Press, Berkeley, 1964. Interesting from both historical and scientific points of view. Brush's introduction and Boltzmann's forewords to Parts I and II provide especially interesting historical sidelights. In the 1890's, when this was written, the fashionable scientific attitude toward the kinetic theory was one of skepticism. Boltzmann was convinced of the basic reality of the theory and published these lectures to give his reasons for this belief. His apparent lack of success so depressed him that he committed suicide in 1906—only a few years before his point of view was completely vindicated by the discovery of quantum theory. Boltzmann's book is not easy to read, but it contains a detailed outline of van der Waals' theory and its difficulties, and is of more than historical interest.

Standard older texts:

J. Jeans, *The Dynamical Theory of Gases*, 4th ed., Dover Publications, New York, 1954 (reprint of the Cambridge Press edition of 1925), Chapter 6.

J. Jeans, *An Introduction to the Kinetic Theory of Gases*, Cambridge, 1940, Chapter 3.

L. B. Loeb, *Kinetic Theory of Gases*, McGraw-Hill, New York, 1927, Chapter 5.

E. H. Kennard, *Kinetic Theory of Gases*, McGraw-Hill, New York, 1938, Chapter 5.

Chapter 3

THE MOLECULAR THEORY OF THE THERMAL ENERGY AND HEAT CAPACITY OF A GAS

3-1 The translational energy of a gas; heat capacity of a monatomic gas

In Eq. (2-17) it was suggested that for n moles of a dilute gas at temperature T it is possible to make the identification

$$nRT = \frac{1}{3} Nm\overline{u^2} \qquad (3\text{-}1)$$

where R is the gas constant, N is the number of molecules present, m is the molecular mass, and $\overline{u^2}$ is the mean square velocity of the molecules. If R is replaced by $N_0 k$, where N_0 is Avogadro's number and k is Boltzmann's constant [Eq. (2-42)], and if N is replaced by nN_0, then Eq. (3-1) can be written

$$kT = \frac{1}{3} m\overline{u^2} \qquad (3\text{-}2)$$

The kinetic energy associated with the motion through space of the center of gravity of an object of mass M at a velocity V is $\frac{1}{2}MV^2$. The energy associated with the motion (translation) of the center of gravity of a molecule is called the *translational energy* of the molecule, in order to distinguish it from other kinds of energy that the molecule might possess. For instance, later in this chapter we shall find that useful results can be obtained if a diatomic molecule such as hydrogen or oxygen is regarded as a pair of massive particles (the atomic nuclei) held together by a spring (the chemical bond). Such an object can undergo two kinds of internal motion: (1) rotation about axes passing through the center of gravity normal to the chemical bond and (2) oscillations of the nuclear masses when the spring-like chemical bond is compressed and stretched. Each of these internal motions makes a contribution to the total energy of the molecule—which will be discussed later in this chapter. For the present we shall confine our attention to the translational energy of gases.

The translational energy of a molecule of mass m moving with velocity u is

$$\epsilon_{\text{trans}} = \frac{1}{2} mu^2 \tag{3-3}$$

In a gas in which the N molecules are given the labels 1, 2, ..., N and in which molecule i has the velocity u_i the total translational energy is

$$E_{\text{trans}} = \frac{1}{2} m \sum_{i=1}^{N} u_i^2 \tag{3-4}$$

Introducing the mean square velocity

$$\overline{u^2} = \frac{1}{N} \sum u_i^2$$

we find

$$E_{\text{trans}} = \frac{1}{2} Nm\overline{u^2} \tag{3-5}$$

From Eqs. (3-1) and (3-2) we see that

$$E_{\text{trans}} = \frac{3}{2} nRT = \frac{3}{2} NkT \tag{3-6}$$

THERMAL ENERGY AND HEAT CAPACITY

and the mean translational energy per molecule is

$$\bar{\epsilon}_{trans} = \frac{3}{2} kT \tag{3-7}$$

The molecules of certain gases are known to consist of a single atom. These *monatomic gases* include the rare gases—He, Ne, A, Kr, Xe, and Rn—and also, provided that the pressure is not too high, the vapors of many metals, such as Hg and Na. According to the quantum theory (to be discussed in more detail below) the internal motions of these atoms (that is, the motions of the electrons in the atoms) are not affected by changes in the temperature of the gas, so long as the temperature is not too high (ordinarily this means temperatures below a few thousand degrees). Therefore the thermal energy of such monatomic gases must be entirely in the form of translational energy. If the symbol E is used for this thermal energy, we can accordingly write

$$E = \frac{3}{2} nRT \tag{3-8}$$

Let a small amount of heat energy, dq, be introduced into a monatomic gas and assume that the process is carried out in such a way that all of this heat increases the thermal energy of the gas molecules, and thus causes the gas to be heated by an amount dT. (This total conversion of dq into thermal energy requires that the system containing the gas do no work on its surroundings on adding the heat. The conditions for assuring this are described in more detail in Volume II of this series.) The increase in thermal energy will be

$$dE = dq = \frac{3}{2} nR \, dT \tag{3-9}$$

The ratio dq/dT is defined as the *heat capacity* of the system and will be denoted by the symbol C_V. (The significance of the subscript V, along with a more general discussion of heat capacities, will be given in Volume II.) From Eq. (3-9) we conclude that

$$C_V = \frac{3}{2} nR \tag{3-10}$$

Since R has a value close to 2 cal/deg mole (see Table 1-2), Eq.

(3-10) leads us to expect that the *molar heat capacities* of all monatomic gases (i.e., the heat capacity when $n = 1$) will be close to $(\frac{3}{2}) \times 2 = 3$ cal/deg mole (a more precise value is $\frac{3}{2} \times 1.987 = 2.980$ cal/deg mole). This value agrees, in fact, with the observed heat capacities of all the rare gases and all other common monatomic gases, to a precision within the experimental error of the heat capacity measurement. This remarkable fact justifies the identification of $\frac{1}{3}N m \overline{u^2}$ with nRT that was made in Eq. (2-17). Thus the Bernoulli theory makes it possible to use the empirical constant R, appearing in the ideal equation of state in order to predict the heat capacity of a gas with high precision. This is an excellent example of the power of the molecular point of view in discovering correlations between apparently quite unrelated properties of real systems.

3-2 *The classical mechanical theory of the heat capacities of diatomic and polyatomic molecules; principle of the equipartition of thermal energy*

In diatomic and polyatomic molecules thermal energy can be absorbed by the internal motions of the molecules (rotations and vibrations) as well as in the form of the kinetic energy of motion of the molecule's center of gravity (translational energy). The relationship between the temperature and the internal energy, and hence the contributions of internal motion to the heat capacity, can be deduced rigorously from classical physics by applying the laws of Newtonian mechanics to the model of a gas as a chaotic collection of moving objects. The details of this theory cannot be given at this point, but it is of interest to consider the theoretical result, which is quite simple and which also happens to be wrong, as we shall see.

It is necessary to introduce the concept of a *mechanical degree of freedom* in a molecule. A degree of freedom is an independent mode of motion, such as a rotation, or a mode of vibration, or a translation in one of the three independent (i.e., mutually perpendicular) directions of space. A more precise account of what is meant by a degree of freedom will be given in Volume II, but for present purposes the meaning will be clear enough if we describe how the number of degrees of freedom is determined. If a molecule contains only one atom, it has only three degrees of freedom, corresponding to translational motions in the three directions of space. If a molecule contains A atoms, then each atom contributes three degrees of freedom, so the molecule has a total of $3A$ degrees of freedom. These $3A$ degrees of freedom are distributed among the following types of motion:

(a) Three degrees of freedom belonging to the translational motion of the center of gravity of the molecule in the three directions of space.

(b) If the atoms of the molecule all lie along a common straight line (as is the case for all diatomic molecules, and also for CO_2, CS_2, N_2O, and acetylene, H—C≡C—H), then only two rotational motions are possible, about the two mutually perpendicular axes that can be drawn normal to the molecular axis. (Since the masses of atoms are concentrated in the nearly pointlike nuclei which lie on the axis, it is presumed to be meaningless to consider rotations about the molecular axis itself.) Thus linear molecules are said to have two rotational degrees of freedom.

For nonlinear polyatomic molecules (e.g., H_2O, SO_2, ethylene, benzene) rotational motion is possible about three mutually perpendicular axes, and there are three rotational degrees of freedom.

(c) The remaining degrees of freedom of the molecule are usually vibrational in character. For linear molecules there are $3A - 5$ of these vibrational modes [$3A - 5 = 3A$ (total) $- 2$(rotational) $- 3$(translational)], whereas for nonlinear molecules there are $3A - 6$ vibrational modes.

For instance, a diatomic molecule ($A = 2$) must be linear, so there are:

3 translational degrees of freedom
2 rotational degrees of freedom
$3 \times 2 - 5 = 1$ vibrational degree of freedom

For carbon dioxide ($A = 3$, linear) we have:

3 translational degrees of freedom
2 rotational degrees of freedom
$3 \times 3 - 5 = 4$ vibrational degrees of freedom

For the water molecule, H_2O ($A = 3$, nonlinear), we have:

3 translational degrees of freedom
3 rotational degrees of freedom
$3 \times 3 - 6 = 3$ vibrational degrees of freedom

It is possible to give a pictorial representation of each of these degrees of freedom as shown in Fig. 3-1. The translational modes are naturally ascribed to overall motion of the center of gravity of the molecule in the three mutually perpendicular directions in space. The rotations are identified with the two (for linear molecules) or three (for nonlinear molecules) possible mutually perpendicular axes of rotation. The various patterns of vibration are deduced from the so-called normal mode analysis of the molecular distortions, which will be discussed elsewhere.[1] In Fig. 3-1 the small arrows indicate the directions of motion of the different atoms at some arbitrary instant during each mode of vibration; the direction of each of these arrows in a given mode is precisely reversed in each half vibration cycle. It is not expected at this point that the student will feel that he fully understands the dynamical principles involved here—though before he has finished his study of physical chemistry he ought to have gone into this subject in some detail. It will be enough if the student grasps the simple classification of degrees of freedom into translational, rotational, and vibrational categories and if he is able to calculate the number of degrees of freedom in each category for any molecule.

We are now in a position to state the classical mechanical rule for predicting the thermal energy of a molecule. This rule is called the *principle of the equipartition of energy*, and states:

(a) each translational and rotational degree of freedom in a molecule contributes $\frac{1}{2}nRT$ to the thermal energy of n moles of a gas;

(b) each vibrational degree of freedom in a molecule contributes nRT to the thermal energy of n moles of a gas—that is, twice as much as the contributions of rotational and translational degrees of freedom.

It is evident that this rule leads to the following expressions for the molar heat capacities of gases containing A atoms per molecule:

[1] See M. W. Hanna, *Quantum Mechanics in Chemistry*, W. A. Benjamin, Inc., New York, 1965, pp. 97–101.

For monatomic gases ($A = 1$)

$$C_V = 3 \cdot \frac{1}{2} R \qquad (3\text{-}11)$$

For linear molecules

$$C_V = 3 \cdot \frac{1}{2} R + 2 \cdot \frac{1}{2} R + (3A - 5)R = \left(3A - \frac{5}{2}\right) R \qquad (3\text{-}12)$$

For nonlinear molecules

$$C_V = 3 \cdot \frac{1}{2} R + 3 \cdot \frac{1}{2} R + (3A - 6)R = (3A - 3)R \qquad (3\text{-}13)$$

In Table 3-1 the values predicted by these formulae are compared with the values of C_V actually observed for various gases at room temperature. It is evident that, except for the monatomic gases, the predicted values deviate greatly from the observed values. The fact that the simple diatomic gases such as H_2, O_2, and N_2 have values of C_V that are close to $\frac{5}{2}R$ at room temperature at first led physicists to overlook the possibility of vibrations in gaseous molecules, the heat capacity coming apparently entirely from translation and rotation. Around 1910, however, it was discovered that below room temperature the heat capacity of hydrogen gas drops rapidly below its value of $\frac{5}{2}R$; just above its boiling point, C_V has fallen to almost $\frac{3}{2}R$. The classical Newtonian theory of mechanics could produce no convincing explanation for this behavior.

The principle of equipartition of energy was derived in various ways by Waterston, Clausius, Maxwell, Boltzmann, Tait, Jeans, and others during the latter half of the 19th century and the first decade of the 20th century. The derivations were all based on the classical laws of mechanics, although additional assumptions had to be introduced having to do with the statistical behavior of molecules. Some scientists (particularly Lord Kelvin) were inclined at the time to take exception to these statistical assumptions, but we now know that the source of the trouble lay elsewhere. The discrepancies between the predicted and observed heat capacities of diatomic and polyatomic gases gradually grew into a source of intense intellectual discomfort. In 1901 Lord Kelvin published a detailed critique entitled "Nineteenth Century Clouds

Molecule	Modes		
	Translational	Rotational	Vibrational
H_2 $A = 2$ $3A = 6$	Motion in plane of page Motion out of plane of page	Axis of rotation in plane of page Axis of rotation normal to plane of page	"Stretch" mode

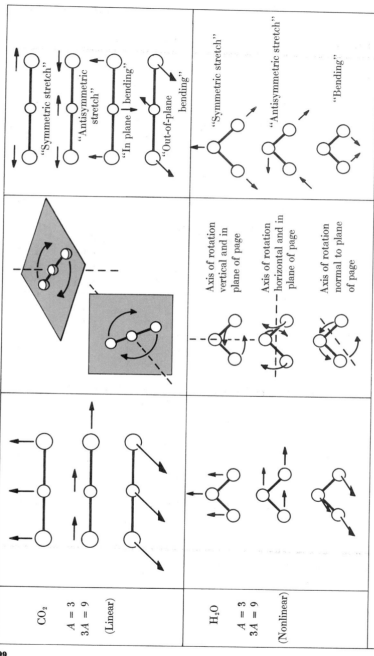

FIG. 3-1 *Modes of motion of some simple molecules.*

TABLE 3-1 **HEAT CAPACITIES OF GASES**

Gas	A^a	C_V from equipartition principle (cal/deg mole)	C_V observed at 25°C (cal/deg mole)	θ in Eq. (3-46) (°K)	τ in Eq. (3-57) (°K)	C_V theory at 25°C (cal/deg mole)
Monatomic gases						
He, Ne, A, Hg, etc.	1	2.98	2.98–3.0	—	—	2.98
Gases with linear molecules; $C_V = (3A - \tfrac{5}{2})R$						
H_2	2	6.95	4.86	5958°	85°	4.87
O_2	2	6.95	4.97	2228°	2.06°	4.97
N_2	2	6.95	4.97	3337°	2.85°	4.97
Cl_2	2	6.95	6.0	797°	0.35°	6.10
Br_2	2	6.95	—	460°	0.12°	6.60
I_2	2	6.95	—	305°	0.05°	6.79
HCl	2	6.95	5.02	4131°	14.9°	4.97
CO_2	3	12.90	6.7	$\begin{Bmatrix}960°(2)\\1930°\\3420°\end{Bmatrix}$	—	6.99
Gases with nonlinear molecules; $C_V = (3A - 3)R$						
H_2O	3	11.90	6.1	$\begin{Bmatrix}2280°\\5220°\\5370°\end{Bmatrix}$	—	6.02
SO_2	3	11.90	7.4	$\begin{Bmatrix}742°\\1650°\\1947°\end{Bmatrix}$	—	7.36
NH_3	4	17.85	6.7	$\begin{Bmatrix}1330°\\2325°(2)\\4770°\\4930°(2)\end{Bmatrix}$	—	6.51

^{a} A = number of atoms per molecule.

over the Dynamical Theory of Heat and Light,"[2] which was largely concerned with the problems raised by the failure of the equipartition principle. By 1910 it had become fairly clear that the trouble was very fundamental indeed, and physicists had to consider the possibility that the Newtonian laws of mechanics are not valid for molecular systems. Classical mechanics at the molecular level then began to be displaced by a new form of mechanics that Planck had started to uncover in 1901.

In order to understand the heat capacities and thermal energies of molecules it is therefore necessary that we consider some of the results of this new form of mechanics.

3-3 Summary of quantum rules for the energies of molecules

According to the quantum theory, and in contradiction to the theory of mechanics developed from Newton's laws, energy cannot be added continuously to most mechanical systems. The systems are required by the basic laws of nature to accept energy in finite increments, or *quanta*. The rules which tell us what energies it is possible for a system to have may be derived from a particular development of quantum theory known as quantum mechanics. In this section we shall summarize these rules for three kinds of systems which are especially important for the understanding of molecular dynamics.[3]

a. PARTICLE IN A BOX   If a particle of mass m moves in a rectangular box of dimensions $a \times b \times c$, its energy can have only the values

$$\epsilon_{\text{trans}} = \frac{h^2}{8m}\left[\left(\frac{n_1}{a}\right)^2 + \left(\frac{n_2}{b}\right)^2 + \left(\frac{n_3}{c}\right)^2\right] \tag{3-14}$$

where h is Planck's constant (6.627×10^{-27} erg sec) and n_1, n_2, and n_3 are positive integers (neither n_1, n_2, nor n_3 can be set equal to zero). The motions of a particle back and forth in a box are the "translational" motions discussed in Section 3-1, so the subscript "trans" has been placed after the symbol for the energy. Each

---

[2] *Phil. Mag.*, **6**, 1(1901).
[3] These quantum rules are derived from the basic postulates of quantum mechanics in another volume of this series, M. W. Hanna, *Quantum Mechanics in Chemistry*, W. A. Benjamin, Inc., New York, 1965.

possible set of three positive integers specifies a possible state of a particle in a box. Equation (3-14) is not at all difficult to derive from the basic principles of quantum theory, and we shall do this in Chapter 4.

The quantization of energy according to Eq. (3-14) had escaped the notice of mankind before about 1910 because the energy levels are extremely close together when one deals with masses of the order of grams and boxes of macroscopic size. For instance, if in Eq. (3-14) we set $m = 1$ g and $a = b = c = 1$ cm, then when $n_1 = n_2 = n_3 = 1$, the energy is 16×10^{-54} erg. If $n_1 = n_2 = 1$ and $n_3 = 2$, the energy is 33×10^{-54} erg. Thus the difference in energy between these two levels is of the order of 10^{-53} erg. It is found that as the quantum numbers n_1, n_2, and n_3 get larger, the quantum levels fall even closer together. These energy differences are, of course, much too small for us to observe directly with our eyes and hands or even with subtle instrumentation, so Eq. (3-14) is consistent with our everyday experience.

b. LINEAR ROTATORS For a linear molecule consisting of two or more atoms arranged along a straight line (e.g., H_2, CO_2, N_2O, acetylene) the energies associated with molecular rotation are given by

$$\epsilon_{\text{rot}} = \left(\frac{h^2}{8\pi^2 I}\right) j(j+1) \qquad (3\text{-}15)$$

where h is again Planck's constant, I is the moment of inertia about the molecular center of gravity, and j is either zero or a positive integer. For each value of the integer j there are $2j + 1$ different states having the same energy but differing in the spatial orientation of the axis of rotation. For a diatomic molecule consisting of two atoms of masses m_1 and m_2 at a distance R apart the moment of inertia is

$$I = \frac{m_1 m_2 R^2}{m_1 + m_2} \qquad (3\text{-}16)$$

Once again, it is easy to see that these energy levels are much too close together in macroscopic systems to be observable: if $m_1 = m_2 = 1$ g and $R = \sqrt{2}$ cm, then $I = 1$ g-cm^2 and $\epsilon_{\text{rot}} = 0.54 \times 10^{-54} j(j+1)$ erg. Even if $\epsilon_{\text{rot}} = 1$ erg, corresponding to a value of j of about 10^{27}, successive levels differ in energy by only about

THERMAL ENERGY AND HEAT CAPACITY

10^{-27} erg. It is no wonder that when we see a wheel slow down we cannot detect that it does so in a series of jumps. The jumps are much too small for us to perceive.

c. OSCILLATORS If any system can undergo harmonic oscillations with a frequency ν cycles per second, then energy is quantized according to the relation

$$\epsilon_{\text{vib}} = vh\nu \qquad (3\text{-}17)[4]$$

where $v = 0, 1, 2, 3, \ldots$ and there is only one possible state of the oscillator for each value of v. This is the quantum rule discovered by Planck in 1901, which began the development of quantum theory. If $\nu = 1$ oscillation per second, the energy quanta are 6.6×10^{-27} erg, which is much too small to be observable with our unaided senses.

d. RELATIVE MAGNITUDES OF ENERGY QUANTA AND THE MEAN THERMAL ENERGY On the atomic scale, the energy differences between successive quantum states may under certain circumstances become comparable with $\frac{1}{2}kT$, the mean thermal energy per degree of freedom available to molecules. This is seen from the following argument. At room temperature (say 300°K), $\frac{1}{2}kT$ is about 2×10^{-14} erg per molecule. This may be compared with the spacing between the rotational energy levels of hydrogen: the mass of the hydrogen atom is 1.66×10^{-24} g, and the distance between the two atoms in the hydrogen molecule is $R = 0.74 \times 10^{-8}$ cm, so that the moment of inertia of H_2 is $I = 0.45 \times 10^{-40}$ g cm^2. Thus $\epsilon_{\text{rot}} = 1.2 \times 10^{-14} j(j+1)$ erg. The energy difference between the lowest rotational state of H_2 (i.e., the state with $j = 0$) and the first excited level (that with $j = 1$) is 2.4×10^{-14} erg, which is of the same order of magnitude as $\frac{1}{2}kT$ at room temperature. Therefore it is not surprising that the classical theory of the rotational thermal energy of hydrogen gas begins to fail when hydrogen is cooled below room temperature. Most other gases do not show this failure because their atoms are heavier than hydrogen, and they are farther apart, so their moments of

---

[4] The quantum rule for vibrational energy is more properly written as

$$\epsilon_{\text{vib}} = \left(v + \frac{1}{2}\right)h\nu \qquad (3\text{-}18)$$

but the contribution of the term $\frac{1}{2}h\nu$ (the zero point energy) has no influence on the thermal properties that are considered in this chapter. See problem 1.

inertia are several times larger and the energy levels are correspondingly closer together. If one tries to cool these gases sufficiently to bring forth the quantum effects on the rotational energy, they condense to the liquid or solid state.

Diatomic molecules—and polyatomic molecules, too—can undergo vibrations which are subject to energy quantization. For diatomic molecules but a single mode of vibration is possible, in which the two atoms move back and forth along the line that joins their centers. This vibration frequency can be measured by studying the ultraviolet, infrared, or Raman spectra of diatomic molecules. For H_2 the frequency is found to be 13.2×10^{13} vibrations per second, so that

$$\epsilon_{\text{vib}} = v \times 13.2 \times 10^{13} \times 6.6 \times 10^{-27}$$
$$= v \times 87 \times 10^{-14} \text{ erg}$$

Thus the energy comes in packets of 87×10^{-14} erg, which is more than 40 times larger than $\frac{1}{2}kT$ at 300°K. The consequence is, as we shall see below, that very few H_2 molecules exist in any but their lowest vibrational state (the state with $v = 0$) and there is practically no thermal vibrational energy in H_2 at room temperature.

The classical theory completely fails here since it predicts a mean thermal vibration energy of kT per hydrogen molecule.

e. THE USE OF QUANTUM NUMBERS TO SPECIFY STATES OF MOLECULES   According to the quantum theory, the state of a diatomic molecule is completely specified if one writes down six quantum numbers. Thus the total energy of a diatomic molecule is

$$\epsilon = \epsilon_{\text{trans}} + \epsilon_{\text{rot}} + \epsilon_{\text{vib}} = \frac{h^2}{8m}\left[\left(\frac{n_1}{a}\right)^2 + \left(\frac{n_2}{b}\right)^2 + \left(\frac{n_3}{c}\right)^2\right] + \left(\frac{h^2}{8\pi^2 I}\right)j(j+1) + vh\nu \quad (3\text{-}19)$$

where m is the mass of the molecule, I is its moment of inertia, and ν is the vibration frequency. This equation contains the five quantum numbers n_1, n_2, n_3, j, and v, and there is an additional quantum number (often called the "azimuthal" or "magnetic" quantum number) which specifies the orientation of the axis of

THERMAL ENERGY AND HEAT CAPACITY 105

rotation, but which has no effect on the energy. The collection of quantum numbers specifies the quantum state of the molecule. A monatomic molecule needs only the three quantum numbers n_1, n_2, n_3 in order to specify its state. In general, if a molecule contains A atoms, then $3A$ quantum numbers are needed in order to specify its state. Linear molecules such as CO_2, H_2, and O_2 have three translational quantum numbers, two rotational quantum numbers, and $3A - 5$ vibrational quantum numbers. (The energy in each vibrational mode of motion of the molecule is quantized according to the Planck rule, Eq. (3-17), the frequency ν being that which corresponds to the particular mode of vibration.) If the molecule is not linear (such as H_2O, CH_4, and benzene), then there are three rotational quantum numbers and $3A - 6$ vibrational quantum numbers. Thus benzene, C_6H_6, with 12 atoms, requires a total of 36 quantum numbers in order to specify the state of the molecule. Three of these numbers give the translational state, three more the rotational state, and the remaining 30 denote the vibrational energy state of the 30 different modes of vibration.

3-4 The average populations of the molecular quantum states in a gas

It is not difficult to insert these notions of quantized energy states into the picture of a gas as a collection of molecules moving through space at random and subject to the laws of probability. The key to the introduction of quantum theory into the problem is to be found in the answer to the following question: "In a gas at temperature T, what is the probability, $P(\epsilon)$, that a molecule will be found in a particular quantum state whose energy is ϵ?" The answer to this question was provided by Ludwig Boltzmann around 1870–1875, many years before the quantum theory was discovered. Boltzmann showed that this probability is proportional to the quantity $\exp(-\epsilon/kT)$. That is

$$P(\epsilon) \propto \exp\left(-\frac{\epsilon}{kT}\right) \qquad (3\text{-}20)$$

where k is Boltzmann's constant (defined in Chapter 2) and e is the base of natural logarithms, $e = 2.71828 \ldots$. The quantity

$\exp(-\epsilon/kT)$ is known as the *Boltzmann factor*. It is probably the most useful combination of concepts in all of physical chemistry and we shall encounter it again and again in this book. In Volume II in this series we shall learn how Boltzmann's factor arises as a direct consequence of the laws of probability. An elementary discussion of the physical basis of the Boltzmann factor is also given in Appendix 3-1.

The Boltzmann factor allows one to calculate the average number of molecules to be found in each quantum state in any given sample of gas. It is convenient to arrange the quantum states of the system in order of increasing energy and to identify each state by a *single* serial number—zero for the lowest state, 1 for the next highest state, 2 for the next one, and so on. (These serial numbers are not to be confused with the quantum numbers—in fact each serial number corresponds to a whole set of $3A$ quantum numbers.) The energy of the state with serial number 0 is ϵ_0, that for the state with serial number 1 is ϵ_1, and so on. Now the average population of a state ought to be proportional to the probability of finding a molecule in that state. Therefore, if $n_0, n_1, n_2, \ldots$ are the average populations of states $0, 1, 2, \ldots$, Boltzmann's result tells us that

$$n_0 = A \exp\left(-\frac{\epsilon_0}{kT}\right)$$

$$n_1 = A \exp\left(-\frac{\epsilon_1}{kT}\right) \quad (3\text{-}21)$$

$$n_2 = A \exp\left(-\frac{\epsilon_2}{kT}\right)$$

. . . .

where A is a proportionality constant whose value is the same for all states. It is easy to find an expression for A. Suppose that the number of molecules in the sample of gas is N. All of the molecules must be in one state or another, so it must be true that

$$N = n_0 + n_1 + n_2 + \cdots = \sum_{i=0}^{\infty} n_i \quad (3\text{-}22)$$

(Since there will generally be an infinite number of states, the serial numbers, i, must be allowed to run to infinity.) This means that

$$N = A\exp\left(-\frac{\epsilon_0}{kT}\right) + A\exp\left(-\frac{\epsilon_1}{kT}\right) + A\exp\left(-\frac{\epsilon_2}{kT}\right)$$
$$+ \cdots = \sum_{i=0}^{\infty} A\exp\left(-\frac{\epsilon_i}{kT}\right) \quad (3\text{-}23)$$

Since A is a constant, it can be factored out of the sum, so that

$$N = A\left(\exp\left(-\frac{\epsilon_0}{kT}\right) + \exp\left(-\frac{\epsilon_1}{kT}\right) + \exp\left(-\frac{\epsilon_2}{kT}\right) + \cdots\right)$$
$$= A\sum_{i=0}^{\infty}\exp\left(-\frac{\epsilon_i}{kT}\right) \quad (3\text{-}24)$$

and

$$A = N \bigg/ \sum_{i=0}^{\infty}\exp\left(-\frac{\epsilon_i}{kT}\right) \quad (3\text{-}25)$$

The sum in the denominator of Eq. (3-25) is a very important quantity which will be encountered repeatedly in this volume as well as in succeeding volumes in the series. It is one of the central concepts in the molecular theory of thermodynamics and is called the *partition function*. It will be denoted by the symbol Q,

$$Q = \sum_{\substack{\text{all}\\\text{states}}} \exp\left(-\frac{\epsilon_i}{kT}\right) \quad (3\text{-}26)$$

Thus we find that the average population of state i is

$$n_i = \left(\frac{N}{Q}\right)\exp\left(-\frac{\epsilon_i}{kT}\right) \quad (3\text{-}27)$$

3-5 *The thermal energy of a system of molecules*

It is clear that the total energy of a collection of molecules must be given by the expression

$$E = n_0\epsilon_0 + n_1\epsilon_1 + n_2\epsilon_2 + \cdots = \sum_{\substack{\text{all}\\\text{states}}} n_i\epsilon_i \quad (3\text{-}28)$$

since ϵ_i is the energy of a single molecule in the ith state and $n_i\epsilon_i$ is the energy of all of the molecules that are in the ith state. This expression may be written

$$E = \left(\frac{N}{Q}\right) \sum_{\substack{\text{all} \\ \text{states}}} \epsilon_i \exp\left(-\frac{\epsilon_i}{kT}\right) = N \frac{\sum \epsilon_i \exp(-\epsilon_i/kT)}{\sum \exp(-\epsilon_i/kT)} \quad (3\text{-}29)$$

where the sums must be taken over all possible quantum states of the molecules.

A direct relationship between E and Q can be obtained in the following way. Note that

$$\frac{d \exp(-\epsilon_i/kT)}{dT} = \left(\frac{\epsilon_i}{kT^2}\right) \exp\left(-\frac{\epsilon_i}{kT}\right) \quad (3\text{-}30)$$

Therefore

$$\epsilon_i \exp\left(-\frac{\epsilon_i}{kT}\right) = kT^2 \frac{d \exp(-\epsilon_i/kT)}{dT} \quad (3\text{-}31)$$

and

$$E = \left(\frac{N}{Q}\right) kT^2 \sum_{\substack{\text{all} \\ \text{states}}} \frac{d \exp(-\epsilon_i/kT)}{dT} = \frac{NkT^2}{Q} \frac{dQ}{dT}$$
$$= NkT^2 \frac{d \ln Q}{dT} \quad (3\text{-}32)$$

For a system containing one mole of a substance, Nk may be replaced by the gas constant, R, and we have

$$E = RT^2 \frac{d \ln Q}{dT} \quad (3\text{-}33)$$

Equations (3-28), (3-29), (3-32), and (3-33) are perfectly general. In particular, nothing in the above derivation restricts these equations to gases. The only assumptions underlying these equations are (1) that molecules exist in quantum states and (2) that the Boltzmann factor is correct.

EXERCISE Prove that if all energy levels are shifted by a constant amount, ϵ_0 (equivalent to choosing a new base from which to measure all energy levels), then the value of the mean thermal energy per molecule, E/N, is also shifted by ϵ_0.

THERMAL ENERGY AND HEAT CAPACITY

Solution All energy levels will now have the energies

$$\epsilon_i' = \epsilon_i + \epsilon_0$$

and Q will now have the form

$$Q' = \sum_i \exp\left(-\frac{\epsilon_i'}{kT}\right) = \sum \exp\left(-\frac{\epsilon_i + \epsilon_0}{kT}\right)$$

$$= \sum \exp\left(-\frac{\epsilon_0}{kT}\right) \exp\left(-\frac{\epsilon_i}{kT}\right)$$

Since the factor $\exp\left(-\frac{\epsilon_0}{kT}\right)$ is the same for all levels we may write

$$Q' = \exp\left(-\frac{\epsilon_0}{kT}\right) \sum \exp\left(-\frac{\epsilon_i}{kT}\right) = \exp\left(-\frac{\epsilon_0}{kT}\right) Q$$

Substitution in Eq. (3-32) gives

$$\frac{E'}{N} = kT^2 \frac{d \ln Q'}{dT} = \frac{E}{N} + \epsilon_0$$

3-6 *The thermal vibrational energy of a diatomic gas molecule*

We are now in a position to calculate how the vibrational energy of a gas with N molecules varies with the temperature—which will make it possible for us to calculate the contributions of vibrational motions to the heat capacities of gases. For a diatomic gas molecule the vibrational energy is

$$\epsilon_v = vh\nu \qquad (v = 0, 1, 2, \ldots)$$

so that the vibrational energy is

$$\begin{aligned}
E_{\text{vib}} &= \epsilon_0 n_0 + \epsilon_1 n_1 + \epsilon_2 n_2 + \cdots \\
&= 0h\nu n_0 + 1h\nu n_1 + 2h\nu n_2 + \cdots \\
&= h\nu \sum_{i=0}^{\infty} i n_i \qquad (3\text{-}34) \\
&= \frac{Nh\nu}{Q_{\text{vib}}} \sum i e^{-ih\nu/kT}
\end{aligned}$$

where

$$Q_{\text{vib}} = \sum e^{-ih\nu/kT}$$
$$= 1 + e^{-h\nu/kT} + e^{-2h\nu/kT} + \cdots \quad (3\text{-}35)$$

Setting $x = e^{-h\nu/kT}$, so that $x^2 = e^{-2h\nu/kT}$, $x^3 = e^{-3h\nu/kT}$, etc., we have

$$Q_{\text{vib}} = 1 + x + x^2 + \cdots = \sum_{i=0}^{\infty} x^i \quad (3\text{-}36)$$

From Newton's formula for the binomial expansion, we find that

$$(1 - x)^{-1} = 1 + x + x^2 + \cdots \quad (3\text{-}37)$$

provided that x lies between -1 and $+1$, which must be the case here. Therefore

$$Q_{\text{vib}} = \frac{1}{1 - x} = \frac{1}{1 - e^{-h\nu/kT}} \quad (3\text{-}38)$$

Furthermore,

$$\sum_{i=0}^{\infty} i e^{-ih\nu/kT} = \sum i x^i = 0x^0 + 1x^1 + 2x^2 + \cdots \quad (3\text{-}39)$$

A simple expression for this sum can be found if we recognize that

$$\frac{d}{dx}(1 + x + x^2 + \cdots) = 0 + 1x^0 + 2x^1 + \cdots \quad (3\text{-}40)$$

and that

$$x\frac{d}{dx}(1 + x + x^2 + \cdots) = x + 2x^2 + 3x^3 + \cdots \quad (3\text{-}41)$$

We may also write

$$x\frac{d}{dx}(1 + x + x^2 + \cdots) = x\frac{d}{dx}\left(\frac{1}{1-x}\right) = \frac{x}{(1-x)^2} \quad (3\text{-}42)$$

so that

$$\sum_{i=0}^{\infty} i x^i = \frac{x}{(1-x)^2} \quad (3\text{-}43)$$

Thus we find

$$E_{\text{vib}} = Nh\nu \frac{x/(1-x)^2}{1/(1-x)} = \frac{Nh\nu x}{1-x}$$

$$= \frac{Nh\nu e^{-h\nu/kT}}{1 - e^{-h\nu/kT}}$$

$$= \frac{Nh\nu}{e^{h\nu/kT} - 1} \tag{3-44}$$

Equation (3-44) can be obtained more directly by using Eq. (3-38) with Eq. (3-32):

$$E_{\text{vib}} = NkT^2 \frac{d \ln Q_{\text{vib}}}{dT}$$

$$= -NkT^2 \frac{d}{dT} \ln(1 - e^{-h\nu/kT})$$

$$= NkT^2 \frac{(h\nu/kT^2)e^{-h\nu/kT}}{1 - e^{-h\nu/kT}}$$

$$= \frac{Nh\nu}{e^{h\nu/kT} - 1} \tag{3-45}$$

It is convenient to define a quantity

$$\theta = \frac{h\nu}{k} \tag{3-46}$$

which has the dimensions of temperature. It is evident that

$$\frac{T}{\theta} = \frac{kT}{h\nu} \tag{3-47}$$

That is, the ratio T/θ is the same as the ratio of the classical thermal energy of an oscillator to the energy quantum, $h\nu$, of the oscillator, and θ is the temperature at which the thermal energy is equal to this energy quantum. Table 3-1 shows values of the constant θ for various diatomic gases, and for different modes of vibration of some polyatomic gases. Equation (3-44) can be written in terms of θ as

$$E_{\text{vib}} = \frac{Nk\theta}{e^{\theta/T} - 1} \tag{3-48}$$

For one mole of oscillators, Nk can be replaced by the gas constant, R, so that

$$E_{\text{vib}} = \frac{R\theta}{e^{\theta/T} - 1} \tag{3-49}$$

When E_{vib} is plotted against T using Eq. (3-44) or (3-49), one obtains the curve shown in Fig. 3-2. At low temperatures, where θ/T is large, the factor $e^{\theta/T}$ becomes very large and

$$\frac{1}{e^{\theta/T} - 1} \cong e^{-\theta/T} \tag{3-50}$$

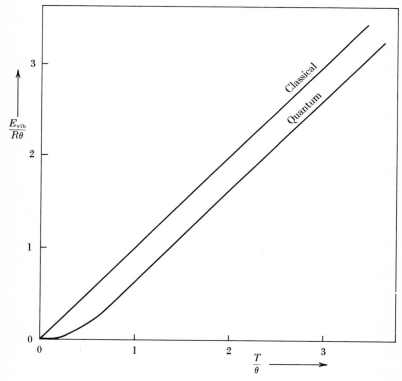

FIG. 3-2 *Temperature dependence of the molar thermal vibrational energy for a classical oscillator (for which $E_{vib} = RT$) and for a quantized oscillator (for which $E_{vib} = R\theta/(e^{\theta/T} - 1)$).*

so that at low temperatures

$$E_{\text{vib}} \cong R\theta e^{-\theta/T} \tag{3-51}$$

and the molar heat capacity is

$$C_{V\text{vib}} = \frac{dE_{\text{vib}}}{dT} \cong R\frac{\theta^2}{T^2} e^{-\theta/T} \tag{3-52}$$

Both the thermal vibrational energy and the contribution of the thermal vibrational energy to the heat capacity are therefore close to zero when $T \ll \theta$. The thermal vibrational energy begins to deviate appreciably from zero when T/θ becomes appreciable relative to unity. Since θ has values of several thousands of degrees Kelvin for H_2, O_2, and N_2 (see Table 3-1) we can understand why the vibrational modes make no appreciable contributions to the heat capacities of these gases at room temperature. For Cl_2, Br_2, and I_2 vapors, θ has the values 797.3°K, 460.4°K, and 305.1°K, respectively, so there should be increasing vibrational contributions to the heat capacities of these gases at room temperature.

As the temperature is raised to very high values, such that $T \gg \theta$, we can expand the exponential factor $e^{\theta/T}$ in a Taylor's series

$$e^{\theta/T} = 1 + \frac{\theta}{T} + \frac{1}{2}\frac{\theta^2}{T^2} + \cdots \tag{3-53}$$

and at sufficiently high temperatures the term $(\theta/T)^2$ can be neglected in comparison with θ/T. We then have

$$E_{\text{vib}} \cong \frac{R\theta}{[1 + (\theta/T) - 1]} = RT \tag{3-54}$$

which is the classical expression for the contribution of a vibrational degree of freedom to the thermal energy. Thus we find that at high temperatures the quantum theory gives the same result as the classical theory. This is, in fact, a general result; at high temperatures the quantum theory and the classical, Newtonian theory tend to give the same value for the thermal energy.

The exact quantum mechanical expression for the contribution of a vibrational degree of freedom to the molar heat capacity is found by differentiating Eq. (3-49)

$$C_{V\text{vib}} = \frac{dE_{\text{vib}}}{dT} = \frac{R(\theta/T)^2 e^{\theta/T}}{(e^{\theta/T} - 1)^2} \tag{3-55}$$

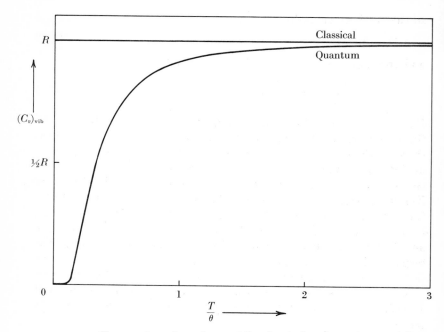

FIG. 3-3 *Temperature dependence of the classical and quantum contributions of a vibrational mode to the molar heat capacity.*

This expression is plotted in Fig. 3-3. Evidently when $T = \theta$, $C_{V\text{vib}} = 0.92R$, so that at the temperature θ, the vibrational contribution to C_V has reached 92% of its classical value. Thus θ is a measure of the temperature below which quantum effects on the thermal vibrational energy become serious.

3-7 The thermal rotational energy of a linear gas molecule

The partition function for the rotational energy of a linear molecule is, from Eqs. (3-15) and (3-26)

$$Q_\text{rot} = \sum_{j=0}^{\infty} (2j+1) e^{-j(j+1)(\tau/T)} \tag{3-56}$$

where the factor $(2j+1)$ before the exponential takes account of the fact that each value of j gives rise to $(2j + 1)$ states having the same energy, and therefore to a group of $(2j + 1)$ identical terms in the partition function. The constant τ in Eq. (3-56) is given by

$$\tau = h^2/8\pi^2 Ik \tag{3-57}$$

I being the moment of inertia of the molecule about its center of mass as defined in Eq. (3-16). Values of τ for various linear molecules are given in Table 3-1.

It is not possible to obtain a simple analytical expression for Q_{rot} which is valid at all temperatures, but it is not difficult to obtain a good approximation to Q_{rot} which is valid at temperatures for which $T \gg \tau$. This is achieved in the following way. Figure 3-4 shows plots of the function, $(2j+1)e^{-j(j+1)(\tau/T)}$ vs. j for various values of T/τ. In the partition function we must know the values of this function only at integral values of j, which are shown as dots along the curves. The numerical value of the sum can be visualized as the sum of the areas of the rectangles in Fig. 3-4, whose height is $(2j+1)e^{-j(j+1)(\tau/T)}$ and whose base is unity. When T/τ is large the sum of the areas of these rectangles will clearly differ only slightly from the area under the smooth curve of $(2j+1)e^{-j(j+1)(\tau/T)}$ vs. j, so it will be a good approximation to write

$$Q_{rot} \cong \int_0^\infty (2j+1)e^{-j(j+1)(\tau/T)}\, dj \tag{3-58}$$

Setting $y = j(j+1)$, and $dy = (2j+1)\, dj$, we see that we can transform this integral to the readily integrated form

$$Q_{rot} \cong \int_0^\infty e^{-(\tau/T)y}\, dy = T/\tau \tag{3-59}$$

Even for small values of T/τ this is not a bad approximation because as can be seen from Fig. 3-4 the deficiency in area between the smooth curve and the rectangles to the left of the maximum term in the partition function will tend to be compensated by the excess area between the smooth curve and the rectangles to the right of the maximum term. In Table 3-2 the exact value of the sum [evaluated by detailed summation of Eq. (3-56) term by term] can be compared with T/τ, which is the approximate value given by Eq. (3-59). From the τ values for typical molecules given in Table 3-1, we find that above room temperature we ordinarily deal with values of T/τ greater than about 150. Only for diatomic molecules containing hydrogen, such as H_2 and HCl, does one find values of T/τ that are appreciably under 100. For H_2, Eq. (3-59) becomes seriously in error somewhat below room temperature.

Substituting Eq. (3-59) into Eq. (3-33) we can readily find the

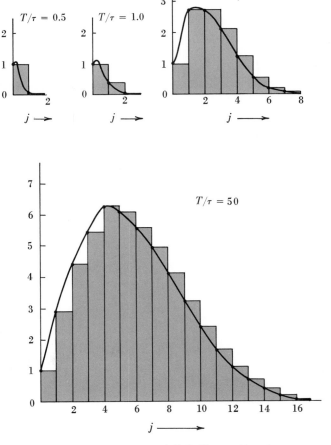

FIG. 3-4 *Plots of* $(2j+1)e^{-j(j+1)\tau/T}$ *(ordinate) against j (abscissa) for various values of T/τ.*

TABLE 3-2 VALUES OF Q_{rot}, E_{rot}, AND $C_{V\,rot}$ FOR VARIOUS VALUES OF THE REDUCED TEMPERATURES, T/τ, WHERE $\tau = h^2/8\pi^2 Ik$

T/τ	Q_{rot}	E_{rot}/RT	$C_{V\,rot}/R$
50	50.007	0.993	1.000
25	25.013	0.987	1.000
10	10.051	0.966	1.000
5	5.348	0.932	1.001
2.5	2.862	0.862	1.005
1.	1.418	0.625	1.073
0.75	1.210	0.466	1.085
0.60	1.108	0.326	0.992
0.5	1.055	0.209	0.794
0.4	1.020	0.0995	0.487
0.3	1.004	0.0256	0.169
0.25	1.001	0.0080	0.064
0.20	1.000	0.0014	0.0137
0.15	1.000	0.0001	0.0009

thermal energy of rotation of one mole of linear molecules at high temperatures,

$$E_{rot} = RT^2 \frac{d \ln Q_{rot}}{dT}$$

$$= RT^2 \frac{d \ln (T/\tau)}{dT} \tag{3-60}$$

$$= RT$$

The contribution of the rotational motions to the molar heat capacity is therefore

$$C_{V\,rot} = R \tag{3-61}$$

Equations (3-60) and (3-61) are the same as the classical expressions for the rotational contributions to the thermal energy and the molar heat capacity of a linear molecule. At low temperatures

($T < \tau$), Q_{rot} becomes independent of T and E_{rot} and $C_{V\,rot}$ both drop to zero. According to Eq. (3-56) Q_{rot} can be considered a function of the dimensionless variable, $x = \tau/T$. Thus

$$\begin{aligned}E_{rot} &= RT^2 \frac{d \ln Q(x)}{dx} \frac{dx}{dT} \\ &= -R\tau \frac{d \ln Q(x)}{dx} \\ &= \frac{R\tau}{Q} \sum (2j+1) j(j+1) e^{-j(j+1)x}\end{aligned} \qquad (3\text{-}62)$$

(Note that Q becomes smaller as x becomes larger, so that $d \ln Q/dx$ is negative.) The heat capacity is

$$\begin{aligned}C_{V\,rot} &= Rx^2 \frac{d^2 \ln Q(x)}{dx^2} \\ &= \frac{Rx^2}{Q^2} (12 e^{-2x} + 180 e^{-6x} + 240 e^{-8x} + 1008 e^{-12x} + \cdots)\end{aligned}$$

(3-63)

The sums in Eqs. (3-62) and (3-63) can be readily evaluated for large values of x (i.e., for low temperatures) and show the behavior of E_{rot} and $C_{V\,rot}$ at low temperatures (see Table 3-2). If x is sufficiently large (roughly $x > 1$), Q can be given the value unity and only the first term need be kept in the sums in Eqs. (3-62) and (3-63). We then find that at low temperatures, E_{rot} and $C_{V\,rot}$ approach zero in the following exponential manner,

$$E_{rot} = 6R\tau e^{-2\tau/T} \qquad (3\text{-}64)$$

$$C_{V\,rot} = 12R(\tau/T)^2 e^{-2\tau/T} \qquad (3\text{-}65)$$

These equations cannot be put to an experimental test because the constant τ is so small for most gases (see Table 3-1). Most substances become liquids with very low vapor pressures at temperatures well above those at which Eqs. (3-64) and (3-65) are valid (i.e., temperatures at which $\tau/T > 1$).

3-8 The thermal translational energy of a gas molecule

The partition function for the translational motion of a gas molecule of mass m moving in a box of dimensions a by b by c is, from Eqs. (3-14) and (3-26)

$$Q_{\text{trans}} = \sum_{n_1=1}^{\infty} \sum_{n_2=1}^{\infty} \sum_{n_3=1}^{\infty} \exp\left\{-\frac{h^2}{8mkT}\left[\left(\frac{n_1}{a}\right)^2 + \left(\frac{n_2}{b}\right)^2 + \left(\frac{n_3}{c}\right)^2\right]\right\} \quad (3\text{-}66)$$

Since the three integers, n_1, n_2, and n_3, must be summed independently, the triple sum can be factored into a product of three sums,[5]

$$Q_{\text{trans}} = \left[\sum_{n_1=1}^{\infty} \exp\left(-\frac{h^2 n_1^2}{8mkTa^2}\right)\right] \left[\sum_{n_2=1}^{\infty} \exp\left(-\frac{h^2 n_2^2}{8mkTb^2}\right)\right] \left[\sum_{n_3=1}^{\infty} \exp\left(-\frac{h^2 n_3^2}{8mkTc^2}\right)\right] \quad (3\text{-}67)$$

[5] The justification for this procedure should be evident from the following examples. Let there be three sets of numbers, a_1 and a_2; b_1, b_2, and b_3; and $c_1, c_2, \ldots, c_N$. Consider first the sum of all products or the a's with the b's:

$$\sum_{j=1}^{3}\sum_{i=1}^{2} a_i b_j = \sum_{j=1}^{3}(a_1 + a_2)b_j = (a_1+a_2)b_1 + (a_1+a_2)b_2 + (a_1+a_2)b_3$$

$$= (a_1+a_2)(b_1+b_2+b_3) = \left(\sum_i a_i\right)\left(\sum_j b_j\right)$$

Next, consider the sum of all products of the a's, b's, and c's:

$$\sum_{k=1}^{N}\sum_{j=1}^{3}\sum_{i=1}^{2} a_i b_j c_k = \sum_k \left(\sum_j \sum_i a_i b_j\right) c_k = \sum_k (a_1+a_2)(b_1+b_2+b_3)c_k$$

$$= (a_1+a_2)(b_1+b_2+b_3)\sum_k c_k$$

$$= \left(\sum_i a_i\right)\left(\sum_j b_j\right)\left(\sum_k c_k\right)$$

Thus the sum of all products formed between the members of different sets of numbers is the same as the product of the sums of the members of the respective sets.

Writing $X = h^2/8mkTa^2$, $Y = h^2/8mkTb^2$, and $Z = h^2/8mkTc^2$ and dropping the subscripts on the three n's, this becomes

$$Q_{trans} = \left(\sum_{n=1}^{\infty} \exp(-Xn^2)\right)\left(\sum_{n=1}^{\infty} \exp(-Yn^2)\right)$$
$$\left(\sum_{n=1}^{\infty} \exp(-Zn^2)\right) \quad (3\text{-}68)$$

It is readily shown that for containers of macroscopic size (a, b, and c all of the order of 1 cm) and for masses of the order of the proton ($m = 1.7 \times 10^{-24}$ g), the constants X, Y, and Z have values of the order of $2.5 \times 10^{-14}/T$. For molecules heavier than the hydrogen atom, the values of X, Y, and Z will be even smaller. Thus at any temperature at which a gas can exist, and for macroscopic containers, the constants X, Y, and Z will be of the order of 10^{-15} or less. For the same reason that the rotational partition function could be approximated by an integral, the sums in Eq. (3-68) can be replaced by integrals (except that here, because of the extremely small values of X, Y, and Z, the error introduced in this way is exceedingly small—very much less than the error involved in evaluating Q_{rot}). Thus we may write

$$\sum_{n=1}^{\infty} \exp(-Xn^2) = \int_0^{\infty} \exp(-Xn^2)\,dn \quad (3\text{-}69)$$

The integral in Eq. (3-69) is the well-known Gauss error integral and has the value $\frac{1}{2}\sqrt{\pi/X}$. By the same reasoning, the two remaining sums in Eq. (3-68) must have the values $\frac{1}{2}\sqrt{\pi/Y}$ and $\frac{1}{2}\sqrt{\pi/Z}$. The translational partition function therefore has the value

$$Q_{trans} = \frac{1}{8}\frac{\pi^{3/2}}{(XYZ)^{1/2}} \quad (3\text{-}70)$$

Substituting for X, Y, and Z we obtain

$$Q_{trans} = \left(\frac{2\pi mkT}{h^2}\right)^{3/2} abc \quad (3\text{-}71)$$

Since $abc = V$, the volume of the container,

$$Q_{\text{trans}} = \left(\frac{2\pi mkT}{h^2}\right)^{3/2} V \qquad (3\text{-}72)$$

Evaluating the molar thermal translational energy from Eq. (3-33) we find

$$E_{\text{trans}} = \frac{RT^2 \, d \ln T^{3/2}}{dT}$$

$$= \frac{3}{2} RT \qquad (3\text{-}73)$$

which is in accord with the classical theory of equipartition and with the results we have already obtained from the kinetic theory of gases [Eq. (3-8)].

3-9 *The complete partition function and the total thermal energy of a diatomic gas*

The energy levels of a diatomic gas molecule are given by the equation (3-19)

$$E = \left(\frac{h^2}{8m}\right)\left[\left(\frac{n_1}{a}\right)^2 + \left(\frac{n_2}{b}\right)^2 + \left(\frac{n_3}{c}\right)^2\right] + \left(\frac{h^2}{8\pi^2 I}\right) j(j+1) + vh\nu \qquad (3\text{-}19)$$

where h is Planck's constant, m, I, and ν are the mass, moment of inertia and vibration frequency of the molecule, a, b, and c are the dimensions of the container of the gas and n_1, n_2, n_3, j, and v are quantum numbers. Recalling that $2j + 1$ states belong to each value of j, we may therefore write for the complete partition function of a diatomic gas molecule

$$Q = \sum_{n_1}\sum_{n_2}\sum_{n_3}\sum_{j}\sum_{v} (2j+1) \exp\left\{-\frac{h^2}{8mkT}\left[\left(\frac{n_1}{a}\right)^2 + \left(\frac{n_2}{b}\right)^2 + \left(\frac{n_3}{c}\right)^2\right] - \left(\frac{h^2}{8\pi^2 IkT}\right) j(j+1) - \left(\frac{vh\nu}{kT}\right)\right\} \qquad (3\text{-}74)$$

This rather frightful expression is easily reduced to a more simple form. The numerical values that any one quantum number may assume are unaffected by the values of the other quantum numbers. Therefore the five summations in Eq. (3-74) may be performed independently. The argument that made it possible to factor the translational partition function, Eq. (3-66), into the form of Eq. (3-67) makes it possible to write Eq. (3-74) in the form

$$Q = \left[\sum_{n_1}\sum_{n_2}\sum_{n_3} \exp\left\{-\frac{h^2}{8mkT}\left[\left(\frac{n_1}{a}\right)^2 + \left(\frac{n_2}{b}\right)^2 + \left(\frac{n_3}{c}\right)^2\right]\right\}\right]$$

$$\cdot \left[\sum_j (2j+1)\exp\left[-\left(\frac{h^2}{8\pi^2 IkT}\right)j(j+1)\right]\right]$$

$$\cdot \left[\sum_v \exp\left(-\frac{vh\nu}{kT}\right)\right] \quad (3\text{-}75)$$

It is evident that the first expression in brackets is the same as Q_{trans} in Eq. (3-66), the second expression in brackets is Q_{rot} in Eq. (3-56) and the third expression in brackets is Q_{vib} in Eq. (3-35). Thus we may write

$$Q = Q_{\text{trans}}Q_{\text{rot}}Q_{\text{vib}} \quad (3\text{-}76)$$

For the total energy of a gas of diatomic molecules we then have

$$E = NkT^2 \frac{d\ln(Q_{\text{trans}}Q_{\text{rot}}Q_{\text{vib}})}{dT}$$

$$= NkT^2 \frac{d\ln Q_{\text{trans}}}{dT} + NkT^2 \frac{d\ln Q_{\text{rot}}}{dT} + NkT^2 \frac{d\ln Q_{\text{vib}}}{dT}$$

$$= E_{\text{trans}} + E_{\text{rot}} + E_{\text{vib}} \quad (3\text{-}77)$$

where E_{trans}, E_{rot}, and E_{vib} are the contributions evaluated in Sections 8, 7, and 6, respectively, of the present chapter. Similarly for the total heat capacity of the gas we may write

$$C_V = C_{V\text{ trans}} + C_{V\text{ rot}} + C_{V\text{ vib}} \quad (3\text{-}78)$$

where $C_{V\text{ trans}}$, $C_{V\text{ rot}}$, and $C_{V\text{ vib}}$ are the contributions of the translational, rotational, and vibrational motions evaluated in the earlier sections of this chapter.

Thus the possibility of writing the energy levels as a sum of independent terms [cf. Eq. (3-19)], each involving a different form of molecular motion, makes it possible to factor the total partition function into a product of partition functions, each determined by a different form of motion. This factorization in turn enables us to express the total energy and heat capacity of a gas as the sum of independent contributions by the different forms of motion. The expressions for E_{trans}, E_{rot} and E_{vib} given in Sections 6, 7, and 8 can therefore be added together to give the total thermal energy of a diatomic gas, and the total heat capacity may be obtained by adding the separate contributions of the translational, rotational and vibrational motions.[6]

For all diatomic gases under practically realizable conditions E_{trans} has its classical value, $3RT/2$, and except for gases which have very low moments of inertia, such as hydrogen—and then only at temperatures below room temperature—E_{rot} has its classical value of RT. For the common diatomic gases, H_2, O_2, N_2, and CO, the vibrational frequency, ν, is so great that the quantity $h\nu/kT$ is of the order of 5 to 10 or more at room temperature. Consequently E_{vib} and $C_{V\,vib}$ are negligibly small for these gases at room temperature. Therefore the quantum theory predicts that for these gases the molar heat capacity at room temperature should be

$$C_V = C_{V\,trans} + C_{V\,rot}$$
$$= \frac{5}{2} R = 4.965 \text{ cal/deg mole} \tag{3-79}$$

Table 3-1 shows that this is observed, so that we may conclude that the failure of the classical equipartition principle has been accounted for by the application of quantum theory.

For diatomic gases at temperatures for which $h\nu/kT$ is of the order of unity or less, quantum theory provides us with a precise formula for the molar heat capacity

$$C_V = \left[\frac{5}{2} + \frac{(\theta/T)^2 e^{\theta/T}}{(e^{\theta/T} - 1)^2} \right] R \tag{3-80}$$

where θ is the quantity defined in Eq. (3-46), $\theta = h\nu/k$.

[6] Strictly speaking, Eq. (3-19) is an approximation because it neglects certain small vibration-rotation interaction terms which are known to be present in real molecules. When these terms are included in Eq. (3-19), the vibrational and rotational contributions to the partition function cannot be factored. It is then no longer possible to separate precisely the contributions of rotations and vibrations to the thermal energy. In practical calculations, however, the corrections to the heat capacity resulting from vibration-rotation interactions are so small that they can almost always be neglected.

For polyatomic gases the classical equipartition theory may be used to find the contributions of translational and rotational motions to the heat capacity, but the quantum theory must be used to find the contributions of each of the vibrational modes. Thus for polyatomic molecules we have

$$C_V = \left[x + \sum_{\substack{\text{all} \\ \text{modes}}} \frac{(\theta_i/T)^2 e^{\theta_i/T}}{(e^{\theta_i/T} - 1)^2} \right] R \qquad (3\text{-}81)$$

where $x = \frac{5}{2}$ for linear molecules and $x = 3$ for nonlinear molecules, and $\theta_i = h\nu_i/k$, where ν_i is the frequency of the ith mode. The constants θ and θ_i in Eqs. (3-80) and (3-81) are generally obtainable with high precision from appropriate spectroscopic measurements of vibration frequencies. These frequencies have been tabulated for many common gases by Herzberg.[7]

Equations (3-80) and (3-81) provide the means of calculating precise heat capacities and thermal energies of many gases over as wide a range of temperatures as these gases are stable. Indeed, since the heat capacities of gases are difficult to measure directly with precision, these equations provide more reliable values of the heat capacities of many gases than are attainable by direct experimental observation.[8] In Table 3-1 are given some values of C_V at 25°C as calculated from Eqs. (3-80) and (3-81).

---

[7] G. Herzberg, *Molecular Spectra and Molecular Structure.* Vol. II: *Infrared and Raman Spectra of Polyatomic Molecules*, Van Nostrand, Princeton, New Jersey, 1945. Molecular vibration frequencies are usually reported in units of cm^{-1}—which gives the number of oscillations in a 1 cm path length of electromagnetic radiation having the same frequency. To convert this number to units of sec^{-1} it is necessary to multiply by the velocity of light. Thus if ω is the frequency in cm^{-1} and ν is the frequency in sec^{-1}, then $h\nu = hc\omega$ where c is the velocity of light, and $\theta = 1.4390\,\omega$°K. Extensive tables of molecular vibration frequencies will also be found in *Landolt-Börnstein, Zahlenwerte und Functionen*, Springer-Verlag, Berlin, 1951, Volume I, Part 2, *Molekeln I*, pp. 227 ff.

[8] The heat capacities of gases are difficult to measure with precision. In one method heat is added at a known rate to a constant stream of gas, and the resulting temperature change of the gas is measured. If Q is the rate of addition of heat (calories per second), R is the flow rate of the gas (moles per second), and ΔT is the temperature change of the gas, then the heat capacity is $Q/R\,\Delta T$. The heat capacity of a gas can also be determined indirectly from the sound velocity by a method to be described in Chapter 2 of Volume II. For a more extensive discussion of the experimental methods of measuring the heat capacities of gases see the references listed at the end of the chapter.

It is necessary to qualify this statement, however, for several classes of molecules. In organic molecules such as ethane and butane, in which there is the possibility of rotation about carbon—carbon single bonds, the energy levels do not obey the simple quantum laws, Eqs. (3-15) and (3-17), because these internal rotational motions are usually neither completely free rotations nor purely harmonic oscillations. For such molecules, Eq. (3-81) is not adequate and additional, more complex terms must be added for the contributions of the internal rotations. Another modification to the theory must be made in the case of hydrogen gas (and also for D_2 and T_2) below room temperature, where nuclear spin effects modify the form of the partition function, Eq. (3-56). This equation is, however, valid for mixed isotopic molecules such as HD. An additional contribution to the thermal energy is also found where excited electronic states are present and when the excitation energy to these states is comparable with kT. Such states are observed for NO (where they make an important contribution to C_V at room temperature) and for O_2 (where the contribution becomes appreciable above about 2000°K).

It must be considered a remarkable accomplishment of the quantum theory and of the molecular theory of heat that they provide these methods for the precise determination of purely thermal properties of matter from spectroscopic data alone. We shall find that these theories also make possible the determination of precise values of other thermodynamic properties from spectroscopic data. Even more important, however, is the insight that these theories provide into the physical basis of thermodynamics and its relationship to molecular structure.

3-10 Simple approximate expressions for the temperature variation of the heat capacity

Although Eqs. (3-80) and (3-81) give a precise description of the temperature variation of the heat capacity of many gases, they are inconvenient for many practical numerical calculations. Therefore these expressions are often replaced in practice by approximate equations which are valid over a limited range of temperature. These equations are usually given the form of a power series fitted to the theoretical curve over a limited range of temperatures. Equations of the following form have frequently been used

$$C_V = a + bT + cT^2 + dT^3 \tag{3-82}$$

$$C_V = a + bT + \frac{c'}{T^2} \tag{3-83}$$

Table 3-3 gives values of the coefficients for some common gases, together with an indication of the range of temperatures for which the equations are applicable. These power series are capable of reproducing the temperature variation of C_V with an error of at

TABLE 3-3 COEFFICIENTS IN THE HEAT CAPACITY EQUATIONS[a]: $C_V = a + bT + cT^2 + dT^3$; $C_V = a + bT + c'T^{-2}$

Gas	a	$10^3 b$	$10^7 c$	$10^9 d$	$10^{-5} c'$	Temp. range (°K)
H_2	4.96	−0.20	4.81	—	—	300–1500
	4.53	0.78	—	—	0.12	300–3000
O_2	4.11	3.25	−10.2	—	—	300–1500
	5.17	1.00	—	—	−0.40	300–3000
N_2	4.46	1.41	−0.81	—	—	300–1500
	4.84	0.90	—	—	−0.12	300–3000
Cl_2	6.78	0.271	—	—	−0.656	300–1500
	6.86	0.16	—	—	−0.68	300–3000
Br_2	6.92	0.14	—	—	−0.0298	300–1500
	6.93	0.12	—	—	−0.30	300–3000
H_2O	5.27	2.298	2.83	—	—	300–1500
	5.31	2.46	—	—	—	300–2750
CO_2	3.165	15.22	−96.8	2.31	—	300–1500
	8.58	2.10	—	—	−2.06	300–2500
SO_2	4.16	13.84	−91.0	2.06	—	300–1800
	9.05	1.88	—	—	−1.84	300–2000

[a] Taken from H. M. Spencer and J. L. Justice, *JACS* **56**, 2311 (1934); H. M. Spencer and G. N. Flannagan, *JACS* **64**, 2511 (1942); H. M. Spencer, *JACS* **67**, 1859 (1945); K. K. Kelley, *U. S. Bur. Mines Bull.* No. 584 (1960). Constants for many other gases will be found in these references. For a given substance the set of constants derived for the more limited temperature range will generally give more accurate heat capacity values in that range.

most a few percent so long as one does not attempt to use them outside the intended temperature range.

The availability of high speed computers makes it possible to utilize the theoretical heat capacity equations directly in computations, so the need for simple approximations such as those described above is decreasing.

3-11 Effects of intermolecular forces on the thermal energy

Throughout this chapter we have assumed that there is no interaction between gas molecules. Thus our formulae for E and C_V apply only to gases at low pressures, conditions under which the ideal gas law is valid. In Volume II we shall show how it is possible to evaluate the effect of gas imperfections on E and C_V from a knowledge of the equation of state. Thus if one has a molecular theory of the equation of state of a nonideal gas, one can deduce the effect of intermolecular forces on the thermal energy and on the heat capacity. The detailed discussion of this matter must, however, be deferred until the student has attained an adequate background in thermodynamics.

Problems

1. Show that if the quantum rule for vibrational energy is written in the form including the zero point energy [cf. Eq. (3-18)]

$$\epsilon = \left(v + \frac{1}{2}\right)h\nu$$

then the vibrational energy of N molecules at temperature T becomes $E'_{\text{vib}} = \frac{1}{2}Nh\nu + E_{\text{vib}}$, where E_{vib} is given by Eq. (3-44). Does the zero point energy make any contribution to C_V?

2. Suppose that some molecule exists in quantum states whose energy is $\epsilon = an^3$, where a is a constant and n is a positive integer. Suppose further that each energy value is nondegenerate (one state for each n). Show that at high temperatures the thermal energy of a gas containing N molecules is $E = \frac{1}{3}NkT$, where k is Boltzmann's constant. [*Hint:* in the integral, $I = \int_0^\infty \exp(-ax^m)\,dx$, set $y = a^{1/m}x$ and note that $I = a^{-1/m}\int_0^\infty \exp(-y^m)\,dy$, where the integral is independent of a.]

3. A molecule can exist in states whose energies are given by $\epsilon = an^\alpha$, where a and α are positive constants and n is a positive integer. Suppose

further that for each value of n there are bn^β states, where b and β are positive constants (i.e., each energy level is "bn^β-fold degenerate"). Show that at high temperatures the thermal energy of a gas containing N of these molecules is $E = [(1 + \beta)/\alpha]NkT$.

4. For the hydrogen atom the electronic energy levels are given by $\epsilon = -R/n^2$ ($n = 1, 2, 3, \ldots$) where R is the Rydberg constant. For each value of n there are $2n^2$ states of the atom. Comparison with Problem 3, above, shows that $\alpha = -2$ and $\beta = 2$, so that the formula given in Problem 3 is $E = -\frac{3}{2}NkT$ for the thermal electronic energy of N hydrogen atoms at high temperatures. Thus, according to the result of Problem 3, the electronic energy of hydrogen atoms should *decrease* (become more negative) as the temperature is raised—which is nonsensical. Can you suggest any reason for this paradoxical result?

5. A system contains N molecules, each capable of existing in three

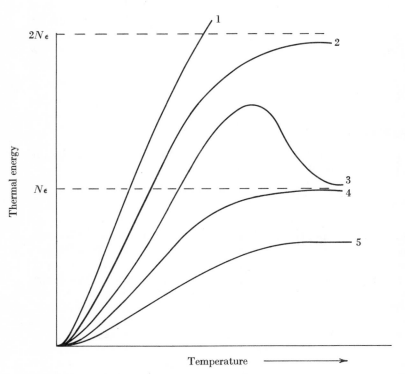

FIG. 3-5 *Conceivable temperature variations of the thermal energy of the system described in Problem 5.*

states, I, II, and III. State I has the energy 0 and states II and III each have energy ϵ. Which of the curves in Fig. 3-5 best approximates the dependence of the thermal energy, E, of the system on the temperature?

6. An atom can exist in two electronic energy states—a ground state of energy 0 and an excited state of energy ϵ. Show that the contribution of this electronic excitation to the molar heat capacity is

$$C_{V\,\text{elect}} = Rx^2 e^x/(1 + e^x)^2$$

where $x = \epsilon/kT$. Draw a rough sketch showing how $C_{V\,\text{elect}}$ varies with T.

7. The lowest electronic energy level of gaseous NO consists of a pair of states. There is also another energy level, consisting of two states, at an energy corresponding to a frequency of 121 cm^{-1} above the lowest level. The bond lengths and bond vibration frequencies are practically identical for all of these states (the vibrational frequency of the N—O bond is 1878 cm^{-1}). (a) Calculate the contribution of the electronic excitation to the heat capacity at 150°K. (b) The observed heat capacity of NO (including translational plus rotational plus vibrational plus electronic energies) at 150°K is $2.75R$. Is this consistent with your answer to part (a)? [Note: See the footnote on page 124 for the conversion of frequencies from cm^{-1} to sec^{-1}.]

8. A gaseous atom can exist in two electronic energy levels—a ground level of energy 0 consisting of a set of g_0 states and an excited level of energy ϵ consisting of a set of g_1 states. Show that the contribution of electronic excitation to the molar heat capacity is

$$C_{V\,\text{elect}} = \frac{RGx^2 e^x}{(G + e^x)^2}$$

where $G = g_1/g_0$ and $x = \epsilon/kT$. Draw rough sketches of $C_{V\,\text{elect}}$ vs. T when $G = 1$, 10 and 100.

9. For Br$_2$ the vibration frequency of the Br—Br bond is 323 cm^{-1} and the Br—Br bond length is 2.28 Å. (a) Find the energy differences (in ergs) between the lowest and the next lowest vibrational energy levels and between the lowest and next lowest rotational levels. (b) Find the two temperatures at which these energy differences become equal to kT.

10. The bond vibration frequency of the oxygen molecule is 1580 cm^{-1}. At what temperature does the vibrational contribution to the heat capacity reach a value of $R/2$?

11. One mole of hydrogen gas is heated from 300°K to 301°K. How much energy (in calories) is absorbed because of population shifts from the vibrational energy state with $v = 0$ to the vibrational state with $v = 1$? The vibration frequency of H$_2$ is 4395 cm^{-1}.

12. Carbonyl sulfide (O=C=S) is a linear molecule with four modes of vibration having the frequencies 527 cm^{-1} (bending mode; doubly degenerate), 859 cm^{-1} and 2079 cm^{-1}. How much energy must be added to one mole of carbonyl sulfide vapor in order to heat it from 25°C to 250°C?

13. Compare the value of C_V for SO_2 at 1000°K obtained from Eq. (3-81) (using values of θ given in Table 3-1) with the values obtained from Eqs. (3-82) and (3-83) (using the constants in Table 3-3).

14. The government of a small country decides to improve the morale of its citizens by distributing a large number, S, of coins at random to its N inhabitants, S being much greater than N. What is the probability that any given citizen will receive s coins? What is the most probable number of coins that a citizen will receive? If $S = 10^8$ and $N = 10^6$, estimate the number of citizens receiving no coins, 100 coins, and 300 coins. (See Appendix 3-1.)

Supplementary references

J. S. Rowlinson, *The Perfect Gas*, Macmillan, New York, 1963. Methods of measuring heat capacities of gases are briefly described in Chapter 2 and detailed comparisons of theoretical and experimental heat capacities for many typical gases are given in Chapter 3. Effects of electronic excitation, nuclear spin and barriers to internal rotation are considered.

J. R. Partington, *Advanced Treatise* (see References, Chapter 1), pp. 792–848. Detailed account of the measurement of heat capacities and a survey of experimental results. Abundant references.

Further discussion of the theory of heat capacities, including the effects of rotational barriers, electronic excitation and nuclear spin, will be found in textbooks of statistical mechanics, such as N. Davidson, *Statistical Mechanics*, McGraw-Hill, New York, 1962, Chapters 8–11 (see also p. 528 for an annotated list of other texts).

Appendix 3-1

PHYSICAL BASIS OF THE BOLTZMANN FACTOR

ALTHOUGH THE Boltzmann factor will be discussed in detail and in more general terms in Volume II, a preliminary justification may be of interest to the student at this point because the factor will be used repeatedly in the rest of this book.

According to the kinetic theory of gases, thermal energy is distributed at random among gas molecules. If the energy of a system is quantized, then there must be a definite (i.e., countable) number of ways of distributing a total energy E among N molecules. In order better to see this, consider a system composed of N identical oscillators whose energy levels obey the quantum rule, Eq. (3-17). The energy in such a system must be distributed among the oscillators in packets of size $h\nu$. Let us, for convenience, express the total energy in terms of the total number of quanta to be distributed, $\mathcal{E} = E/h\nu$. We should like to find the number, $\mathfrak{N}(\mathcal{E}, N)$, of ways in which the $\mathcal{E}$ quanta can be distributed among the N oscillators. This number is the same as the number of ways of distributing ε identical objects among N different boxes, and it may be easily determined by considering the following simple device. Arrange $\varepsilon + N$ dots in a row; there will, of course, be $\varepsilon + N - 1$ spaces between these dots. If lines are drawn in $N - 1$ of the spaces between the dots, with no more than one line being assigned to each space, the ε remaining unfilled spaces will be separated by the lines into N groups. If each of these groups is assigned to a particular box, then we see that each way of drawing $N - 1$ lines in the $\varepsilon + N - 1$ spaces corresponds to one way of distributing the ε objects among the N boxes. For instance, consider the distribution of 5 identical objects among three boxes, A, B, and C ($\varepsilon = 5$, $N = 3$). We draw eight dots in a row and draw two lines ($= N - 1$) in the seven spaces as follows

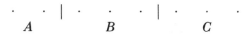

We identify the number of spaces (one) to the left of the first line with the number of objects that might have been placed in box A in one of the distributions. The number of objects placed in boxes B and C are identified with the number of spaces, respectively, between the first and second lines and to the right of the second line. Thus the above arrangement of lines corresponds to a distribution in which one object is placed in box A and two are placed in each of B and C. Similarly, the arrangement

corresponds to a distribution in which no objects are placed in boxes A and B and five are placed in C.

Thus the number $\mathfrak{N}(\varepsilon, N)$ which we seek is the same as the number of ways of drawing $N - 1$ lines in $\varepsilon + N - 1$ spaces. For the example $\varepsilon = 5$, $N = 3$, two lines must be drawn in seven spaces. The first of these two lines can go into any of the seven spaces, and the second line can go into any of the remaining six unfilled spaces. One might at first think that this would indicate that there are $7 \times 6 = 42$ ways of drawing the lines, but we must not overlook the fact that, by counting the number of arrangements in this way we have, for instance, effectively considered the following two arrangements as corresponding to two different

distributions,

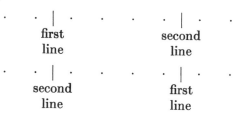

whereas, according to the identifications that we have been using, both ways of drawing the lines correspond to the same distribution. Thus the product, 7 × 6, overcounts the number of distributions by a factor of two; there are actually $\frac{1}{2}(7 \times 6) = 21$ ways of distributing five identical objects in three boxes. [Writing the numbers of objects in A, B, and C for each distribution as (n_A, n_B, n_C), these distributions are: (5, 0, 0); (0, 5, 0); (0, 0, 5); (4, 1, 0); (4, 0, 1); (1, 4, 0); (0, 4, 1); (1, 0, 4); (0, 1, 4); (3, 2, 0); (3, 0, 2); (2, 3, 0); (0, 3, 2); (2, 0, 3); (0, 2, 3); (3, 1, 1); (1, 3, 1); (1, 1, 3); (2, 2, 1); (2, 1, 2); (1, 2, 2).]

In general we can see that

$$\mathfrak{N}(\mathcal{E}, N) = \frac{(\mathcal{E} + N - 1)(\mathcal{E} + N - 2)(\mathcal{E} + N - 3) \cdots (\mathcal{E} + 1)}{(N - 1)!} \qquad (3\text{-}1\text{-}1)$$

The factors in the numerator arise because the "first line" can be drawn in any of $(\mathcal{E} + N - 1)$ spaces, the "second line" can go into any of the remaining $(\mathcal{E} + N - 2)$ spaces, and so on, with the "$(N - 1)$st line" going into any of $(\mathcal{E} + 1)$ spaces. The quantity $(N - 1)!$ in the denominator arises because the product in the numerator counts each of $(N - 1)!$ permutations of $N - 1$ lines among a given set of spaces as giving rise to a different distribution, whereas all of these permutations correspond, in fact, to the same distribution.

Equation (3-1) can also be written as

$$\mathfrak{N}(\mathcal{E}, N) = \frac{(\mathcal{E} + N - 1)!}{\mathcal{E}!(N - 1)!} \qquad (3\text{-}1\text{-}2)$$

Let us now suppose that there are many more quanta to be distributed than there are molecules ($\mathcal{E} \gg N$). Then it will be a reasonable approximation to replace each of the $N - 1$ factors

$(\mathcal{E} + N - j)$ in the numerator of Eq. (1) by $\mathcal{E}$ and writing

$$(\mathcal{E} + N - 1)(\mathcal{E} + N - 2)(\mathcal{E} + N - 3) \cdots (\mathcal{E} + 1) \cong \mathcal{E}^{N-1} \tag{3-1-3}$$

Thus we obtain the approximate relation, valid if $\mathcal{E} \gg N$,

$$\mathfrak{N}(\mathcal{E}, N) = \frac{\mathcal{E}^{N-1}}{(N-1)!} \tag{3-1-4}$$

As an indication of the validity of this approximation, we find that for $\mathcal{E} = 90$ and $N = 10$ the precise value of $\mathfrak{N}$ is $10^{12.24}$ whereas the approximate equation gives $10^{12.03}$. For larger values of $\mathcal{E}/N$ the agreement would be even better.

We are now in a position to determine the probability of finding a particular oscillator in a state with energy ϵ when the group of N oscillators of which it is a member has a total energy $\mathcal{E}$. The laws of probability tell us that the probability of an event is proportional to the number of equivalent ways in which the event can occur. Thus the probability, $P(\epsilon)$, of finding a particular oscillator in a state of energy ϵ is proportional to the number of ways of distributing the remaining energy, $\mathcal{E} - \epsilon$, among the remaining $N - 1$ oscillators, or, assuming that $\mathcal{E} \gg N$,

$$P(\epsilon) \propto \mathfrak{N}(\mathcal{E} - \epsilon, N - 1) = \frac{(\mathcal{E} - \epsilon)^{N-2}}{(N-2)!} \tag{3-1-5}$$

(Note that we have here made the reasonable assumption that all ways of distributing the energy $\mathcal{E} - \epsilon$ among the $N - 1$ oscillators are "equivalent.") Equation (3-5) can be rearranged to give

$$P(\epsilon) \propto \frac{(\mathcal{E} - \epsilon)^{-2}}{(N-2)!} \mathcal{E}^{N} \left(1 - \frac{\epsilon}{\mathcal{E}}\right)^{N} \tag{3-1-6}$$

Let us define a mean number of quanta per molecule,

$$\bar{\epsilon} = \frac{\mathcal{E}}{N} \tag{3-1-7}$$

Assume that only a small fraction of the total energy is given to the particular oscillator, so that $\epsilon \ll \mathcal{E}$. The factor $(\mathcal{E} - \epsilon)^{-2}$ in

PHYSICAL BASIS OF THE BOLTZMANN FACTOR

Eq. (3-6) can be replaced by $\bar{\varepsilon}^{-2}$ and we have

$$P(\epsilon) \propto \frac{\bar{\varepsilon}^{N-2}}{(N-2)!}\left(1 - \frac{\epsilon}{N\bar{\epsilon}}\right)^N \tag{3-1-8}$$

Since N and $\bar{\varepsilon}$ are constants independent of ϵ and since we are concerned only with the dependence of $P(\epsilon)$ on ϵ, we may drop the first factor on the right in Eq. (3-8) and write

$$P(\epsilon) \propto \left(1 - \frac{\epsilon}{N\bar{\epsilon}}\right)^N \tag{3-1-9}$$

Now it is shown in courses in the calculus that

$$\lim_{N \to \infty}\left(1 - \frac{x}{N}\right)^N = e^{-x} \tag{3-1-10}$$

(This may be verified by comparing the binomial expansion

$$\left(1 - \frac{x}{N}\right)^N = 1 - N\frac{x}{N} + \frac{N(N-1)}{2!N^2}x^2$$
$$- \frac{N(N-1)(N-2)}{3!N^3}x^3 + \cdots \tag{3-1-11}$$

with the series that defines e^{-x}

$$e^{-x} = 1 - x + \frac{1}{2!}x^2 - \frac{1}{3!}x^3 + \cdots \tag{3-1-12}$$

If N is large we have to a good approximation $N(N-1)/N^2 \cong 1$, $N(N-1)(N-2)/N^3 \cong 1$, etc., so that the right-hand sides of Eqs. (3-11) and (3-12) become identical.)

Thus we may conclude that in dealing with systems containing very large numbers of oscillators ($N \gg 1$),

$$P(\epsilon) \propto \exp\left(-\frac{\epsilon}{\bar{\epsilon}}\right) \tag{3-1-13}$$

If the mean thermal vibrational energy, $\bar{\epsilon}$, is equated to its classical value, kT, we obtain

$$P(\epsilon) \propto \exp\left(-\frac{\epsilon}{kT}\right) \tag{3-1-14}$$

which is Eq. (3-20).

The physical basis for the Boltzmann factor should be quite clear from this discussion, even though it has been restricted to a collection of oscillators—which have a particularly simple energy quantization rule. The Boltzmann factor arises because the probability that a particular molecule will have a given amount of energy is proportional to the number of ways in which the remaining molecules can distribute the remaining energy. This number decreases very strongly when any one molecule is given a great deal more than the average energy per molecule, $\bar{\epsilon}$. (In Problem 10 of Chapter 4 the Boltzmann factor is derived for a collection of particles in a box. More general derivations will be given in Volume II.)

Chapter 4

THE DISTRIBUTION OF MOLECULAR VELOCITIES IN A GAS

According to the Bernoulli model of a gas, the molecules of a gas move randomly in space. If this is true, then despite the fact that the mean molecular velocity is determined by the temperature, we must expect that individual molecules may at any instant move with quite different velocities. In this chapter we shall show how this velocity distribution can be deduced from the laws of probability, the laws of mechanics and the Boltzmann factor.

4-1 The concept of a distribution function

The problem of quantitatively describing the fact that molecules in a gas move with different velocities is a statistical one, and there are several ways of dealing with it.

Let $f(u)$ be the fraction of the molecules in a given sample of gas whose velocities are less than u. Figure 4-1 shows the general character that one would expect to find for a plot of $f(u)$ against u. Since u is, by definition, a positive quantity (we are not at this point concerned with the directions in which molecules move), and since $u = 0$ is the smallest velocity that a molecule can have, $f(u)$ must be zero when u is zero. Thus the plot of $f(u)$ against u must pass through the origin. On the other hand, relatively few molecules have very large velocities, so that $f(u)$ must approach unity asymptotically as u becomes very large. It is therefore self-evident that $f(u)$ must have something like the general shape

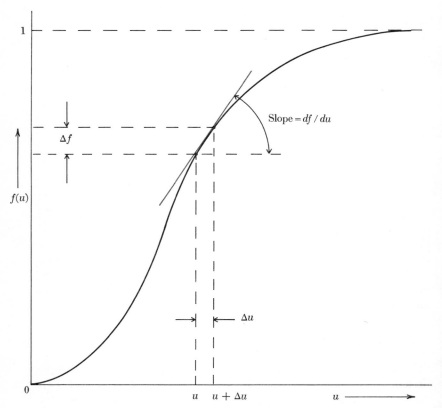

FIG. 4-1 *Dependence on velocity, u, of the fraction, $f(u)$, of molecules in a gas whose velocity is less than u.*

shown in Fig. 4-1. Note that $f(u)$ is numerically equal to the *probability* that a given molecule has a velocity somewhere between 0 and u.

The function $f(u)$ does not turn out to be particularly useful for many applications, however. More interesting is the probability of finding a molecule (or the fraction of the molecules) having a velocity lying somewhere in the range between two particular values, u and $u + \Delta u$. If Δu is small, this probability or fraction is given by (cf. Fig. 4-1)

$$\Delta f = \frac{df(u)}{du} \Delta u \qquad (4\text{-}1)$$

It is convenient to write

$$P(u) = \frac{df}{du} \qquad (4\text{-}2)$$

The new function $P(u)$ is called the *distribution function for molecular velocities*. This function, which is the slope of the curve in Fig. 4-1, is plotted in Fig. 4-2. As drawn it goes through a maximum; nothing that has been stated up to this point about the general character of $f(u)$ requires that this be the true shape of $P(u)$ for a real gas, but we shall find that the shape shown in Fig. 4-2 is indeed the actual shape of $P(u)$ in gases.

Distribution functions such as $P(u)$ are utilized in most statistical problems which involve a continuously varying quantity such as the velocity u. It is important to understand what they mean. Since u is a continuous variable, it is meaningless to speak of "the probability that a molecule has precisely the velocity u." This probability is, in fact, zero, because each molecule can have an infinite number of velocities, even in a limited range of values of u, so that the probability that the molecule will be found with exactly one particular value of u must be vanishingly small. A finite probability for a velocity can be defined only if one gives the molecule the possibility of a finite range of velocities, Δu, in the vicinity of a given velocity, u. If the range Δu is small, then this probability must be proportional to the magnitude of Δu; doubling the range, Δu, will double the probability of finding a molecule whose velocity is in that range. Thus $P(u)$ can be regarded as the proportionality factor relating the range to the probability:

$$\left.\begin{array}{l}\text{Probability that a}\\\text{molecule has a velocity}\\\text{between } u \text{ and } u + \Delta u\end{array}\right\} = P(u)\,\Delta u \quad (4\text{-}3)$$

We shall now show that for a gas at temperature T consisting of molecules of mass m

$$P(u) = Au^2 \exp\left(-\frac{mu^2}{2kT}\right) \quad (4\text{-}4)$$

where A is a quantity independent of u, but dependent on m and T, and where k is Boltzmann's constant (Chapter 2, Section 1d). This is the *Maxwell-Boltzmann distribution function*. Evidently it consists of two factors: (1) a factor Au^2, which increases monotonically as u increases, and (2) a Boltzmann factor $\exp(-mu^2/2kT)$, which decreases monotonically from 1 at $u = 0$ to zero as u

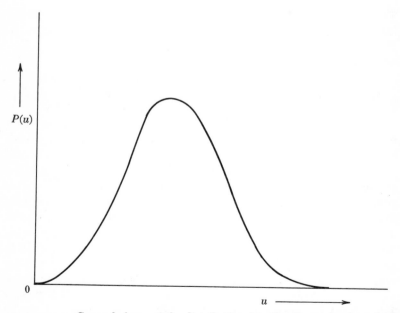

FIG. 4-2 *General shape of the distribution function for molecular velocities, $P(u) = df/du$, where $f(u)$ is the function plotted in Fig. 4-1.*

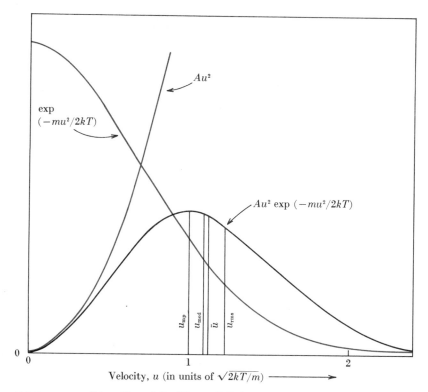

FIG. 4-3 Dissection of the Maxwell-Boltzmann distribution function into the factors Au^2 and $\exp(-mu^2/2kT)$.

increases. The behavior of these two factors and of their product, $P(u)$ is shown in Fig. 4-3.

Although the Maxwell-Boltzmann distribution was originally derived using classical mechanical concepts, it is somewhat easier to derive it if one approaches the problem from a quantum theoretical point of view. Furthermore, some of the mathematical procedures employed in this derivation will prove useful in discussing other subjects. Before proceeding with the derivation, however, it will be helpful to consider some of the quantum principles governing the motions of free particles in closed regions of space.

4-2 Velocity quantization for particles in boxes

In Section 3-3a it was stated that the kinetic energy of a particle of mass m moving back and forth in a box of dimensions a by b by c is quantized according to the law

$$E = \frac{h^2}{8m}\left[\left(\frac{n_1}{a}\right)^2 + \left(\frac{n_2}{b}\right)^2 + \left(\frac{n_3}{c}\right)^2\right] \tag{4-5}$$

where h is Planck's constant and n_1, n_2, and n_3 are positive, non-zero integers. It is of interest to investigate the physical basis of this quantum rule.[1]

According to de Broglie, a particle of mass m moving with velocity u has associated with it a wave whose wavelength is

$$\lambda = \frac{h}{mu} \tag{4-6}$$

Furthermore, de Broglie hypothesized that energy quantization results from the requirement that particles exist in states for which the wave associated with the particles' motion is a standing wave.

For instance, if a particle moves in one dimension in a restricted region of space of length a (e.g., a bead bouncing back and forth without friction on a wire between two stops set at a distance a from each other), then waves travel back and forth with the particle in a manner analogous to the waves that can move up and down a stretched string. These waves can interfere with each other to form a standing wave in the region a only if their wavelength, λ, satisfies the condition

$$n\lambda = 2a \tag{4-7}$$

where n is a positive integer. (This condition is, of course, the exact analogue of the standing wave condition for, say, a violin string of length a.[2]) Substituting Eq. (4-7) in Eq. (4-6) we find that a mass m moving in a one dimensional "box" of length a can

---

[1] A more detailed derivation of this relationship will be found in M. W. Hanna, *Quantum Mechanics in Chemistry*, W. A. Benjamin, Inc., New York, 1965, pp. 45–57.

[2] The displacement, $f(x, t)$, at a point x and a time t in a standing wave in a one dimensional system can be represented by a wave of the general shape

$$f(x, t) = \left[A \sin\left(\frac{2\pi x}{\lambda}\right) + B \cos\left(\frac{2\pi x}{\lambda}\right)\right] \sin 2\pi \nu t$$

where A and B are constants, λ is the wavelength and ν is the frequency of only have velocities

$$u = \frac{\pm nh}{2ma} \tag{4-8}$$

the plus and minus signs arising from the possibility that the molecule can move in two directions. Since the kinetic energy is $E = mu^2/2$, this means that the energy is restricted to the values

$$E = \frac{n^2h^2}{8ma^2} \tag{4-9}$$

Thus de Broglie's hypotheses can be seen to lead to the concept of energy quantization. Quantization is evidently a consequence of some wavelike phenomenon associated with the motion of matter. Furthermore, if a particle moves with constant velocity in a restricted region of space, then the velocity, too, must be quantized.

If a particle moves freely in a two-dimensional box of dimensions a by b (e.g., a nonrotating ball moving with no friction on a pool table), then analysis of the comparable vibrational problem in an elastic membrane reveals that standing waves are possible only if the projections λ_x and λ_y of the de Broglie wave on the side of the box (see Fig. 4-4) satisfy the conditions [3]

$$n_1 \lambda_x = 2a \qquad n_2 \lambda_y = 2b \tag{4-10}$$

---

the standing wave. The external constraints acting on the system (e.g., the fact that the ends of the string are rigidly fixed) require that $f(x, t) = 0$ at both ends of the system at all times (i.e., at $x = 0$ and at $x = a$ for all t). The condition $f(0, t) = 0$ requires that $B = 0$ and the condition $f(a, t) = 0$ requires that $\sin 2\pi a/\lambda = 0$. Since $\sin n\pi = 0$ if and only if n is an integer we may write $2\pi a/\lambda = n\pi$ and obtain Eq. (4-7). The lengths a and λ being inherently positive quantities, only positive values of n can have physical significance.

[3] Consider a rectangular membrane of dimensions a by b cm whose edges are clamped in a fixed position. Let the coordinate axes be located along two of the fixed edges ($x = 0$ and $y = 0$) so that the other two edges are at $x = a$ and $y = b$. If the membrane is vibrating in a standing wave pattern, the displacement $f(x, y, t)$ at point (x, y) and time t is

$$f(x, y, t) = A \sin \frac{2\pi x}{\lambda_x} \sin \frac{2\pi y}{\lambda_y} \sin 2\pi \nu t$$

where λ_x and λ_y are the wavelength projections defined above and ν is the frequency. This function satisfies the requirement that the edges at $x = 0$ and $y = 0$ do not move. The requirement that the edges at $x = a$ and $y = b$ are stationary leads to the conditions $\sin (2\pi a/\lambda_x) = 0$ and $\sin (2\pi b/\lambda_y) = 0$ from which Eqs. (4-10) are obtained.

where n_1 and n_2 are positive integers. The projections λ_x and λ_y are related to the angles α and β between the direction of motion and the sides a and b by (cf. Fig. 4-4)

$$\lambda_x = \frac{\lambda}{\cos \alpha} \qquad \lambda_y = \frac{\lambda}{\cos \beta} \tag{4-11}$$

where λ is given by Eq. (4-6). Furthermore, the components u_x

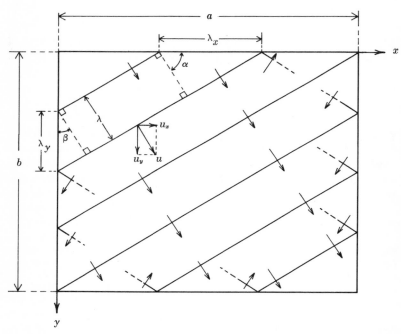

FIG. 4-4 (a) *Motion of a portion of a wave system (wavelength λ) across a two-dimensional rectangular surface of dimensions a by b. The wave front shown here represents a set of wave troughs moving from the upper left toward the lower right, making angles α and β with sides a and b. The projection of the trough-to-trough distance on side a is $\lambda_x = \lambda/\cos \alpha$ and the projection of this distance on side b is $\lambda_y = \lambda/\cos \beta$. Where each wave intersects an edge of the surface there is a reflected wave, shown here only in part. Note that on reflection at an edge a trough becomes a crest and vice versa. The entire system of moving waves plus their reflections will create a standing wave pattern if the ratios a/λ_x and b/λ_y are both integers. In the example shown here $a/\lambda_x = 3$ and $b/\lambda_y = 4$.*

DISTRIBUTION OF MOLECULAR VELOCITIES

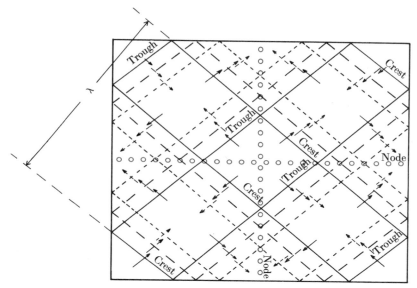

FIG. 4-4 (b) *Generation of a standing wave pattern by the complete wave system when $a/\lambda_x = 2$, $b/\lambda_y = 2$. Successive positions in time of the wave are shown by solid lines, dashed lines, and dotted lines. There are two nodal lines, shown by the lines of open circles, along which displacements caused by the various intersecting waves cancel at all times. Note that because of the change in phase on reflection the lines sloping from upper left to lower right represent wave crests whereas those sloping from upper right to lower left represent wave troughs. (Thus crests and troughs intersect at all times along the nodal lines.)*

and u_y of the velocity of the particle, u, parallel to a and b are given by [4]

$$u_x = \pm u \cos \alpha \qquad u_y = \pm u \cos \beta \qquad (4\text{-}12)$$

Combining Eqs. (4-6), (4-11), and (4-12), we may write

$$\lambda_x = \frac{h}{mu_x} \qquad \lambda_y = \frac{h}{mu_y} \qquad (4\text{-}13)$$

[4] Note that for a given value of α the particle may move in either the positive or negative x-directions, and similarly, for a given β the motion can be in the positive or negative y-directions.

which on substitution in Eq. (4-10) shows us that u_x and u_y must each be quantized with the values

$$u_x = \frac{\pm n_1 h}{2ma} \qquad u_y = \frac{\pm n_2 h}{2mb} \tag{4-14}$$

Furthermore, since

$$E = \frac{1}{2} mu^2 = \frac{1}{2} m(u_x^2 + u_y^2) \tag{4-15}$$

the energy associated with motion in two dimensions must also be quantized, because on substitution of Eq. (4-14) into Eq. (4-15) we find

$$\begin{aligned} E &= \frac{1}{2} m \left[\left(\frac{n_1 h}{2ma} \right)^2 + \left(\frac{n_2 h}{2mb} \right)^2 \right] \\ &= \frac{h^2}{8m} \left[\left(\frac{n_1}{a} \right)^2 + \left(\frac{n_2}{b} \right)^2 \right] \end{aligned} \tag{4-16}$$

For the three-dimensional case, assume that a particle moves in a rectangular box with dimensions a by b by c with velocity u in a direction making angles α, β, γ with the edges a, b, and c, respectively. By analogy with the two-dimensional case, standing waves will be possible only if

$$n_1 \lambda_x = 2a \qquad n_2 \lambda_y = 2b \qquad n_3 \lambda_z = 2c \tag{4-17}$$

where n_1, n_2, and n_3 are positive integers and where

$$\lambda_x = \frac{\lambda}{\cos \alpha} \qquad \lambda_y = \frac{\lambda}{\cos \beta} \qquad \lambda_z = \frac{\lambda}{\cos \gamma} \tag{4-18}$$

Since the components of velocity parallel to the three edges are

$$u_x = \pm u \cos \alpha \qquad u_y = \pm u \cos \beta \qquad u_z = \pm u \cos \gamma \tag{4-19}$$

the de Broglie standing wave condition restricts the velocity components of the particle to the values

$$u_x = \frac{\pm n_1 h}{2ma} \qquad u_y = \frac{\pm n_2 h}{2mb} \qquad u_z = \frac{\pm n_3 h}{2mc} \tag{4-20}$$

Furthermore, since

$$u^2 = u_x^2 + u_y^2 + u_z^2 \tag{4-21}$$

the possible values of the velocity are restricted by the condition

$$u^2 = \frac{h^2}{4m^2}\left[\left(\frac{n_1}{a}\right)^2 + \left(\frac{n_2}{b}\right)^2 + \left(\frac{n_3}{c}\right)^2\right] \quad (4\text{-}22)$$

The energy quantization rule, Eq. (4-5), is obtained from this by setting $E = mu^2/2$.

4-3 Derivation of the Maxwell-Boltzmann distribution function

The evaluation of the probability, $P(u)\,\Delta u$, of finding a molecule with a velocity in the range between u and $u + \Delta u$ may be accomplished by considering the total probability that a particle will be found in any quantum state belonging in this range. Because of the Boltzmann factor (see Chapter 3) the probability that a particle will be found in a particular state with the quantum numbers n_1, n_2, n_3 is

$$P(n_1, n_2, n_3) = \left(\frac{1}{Q}\right)\exp\left[-\frac{E(n_1, n_2, n_3)}{kT}\right] \quad (4\text{-}23)$$

where Q is the translational partition function

$$Q = \left(\frac{2\pi mkT}{h^2}\right)^{3/2} abc \quad (4\text{-}24)$$

$E(n_1, n_2, n_3)$ is the energy of the state, which is also given by $E = \tfrac{1}{2}mu^2$. Thus Eq. (4-23) may also be written

$$P(n_1, n_2, n_3) = \frac{1}{Q}\exp\left(-\frac{mu^2}{2kT}\right) \quad (4\text{-}25)$$

The probability, $P(u)\,\Delta u$, of finding a particle in any state having a velocity in the range u to $u + \Delta u$ will, by the laws of probability, be

$$P(u)\,\Delta u = \sum_{u, u + \Delta u} P(n_1, n_2, n_3) \quad (4\text{-}26)$$

where the sum is taken over all states (i.e., over all combinations of n_1, n_2, and n_3) that lead to velocities between u and $u + \Delta u$. If we require that $\Delta u \ll u$, then the Boltzmann factors in Eq. (4-26) will be essentially the same for all states in the sum and

Eq. (4-26) may be written

$$P(u)\,\Delta u = S(u,\,\Delta u)\frac{1}{Q}\exp\left(-\frac{mu^2}{2kT}\right) \quad (4\text{-}27)$$

where $S(u, \Delta u)$ is the number of combinations of values of n_1, n_2, and n_3 that are consistent with a velocity lying in the range u to $u + \Delta u$.

The quantity $S(u, \Delta u)$ is not difficult to evaluate. Equation (4-22) can be written in the form

$$1 = \left(\frac{n_1}{N_x}\right)^2 + \left(\frac{n_2}{N_y}\right)^2 + \left(\frac{n_3}{N_z}\right)^2 \quad (4\text{-}28)$$

where we write

$$N_x = \frac{2mau}{h} \quad N_y = \frac{2mbu}{h} \quad N_z = \frac{2mcu}{h} \quad (4\text{-}29)$$

It is useful to define a three-dimensional "quantum number space" in which n_1, n_2, and n_3 are plotted along three mutually perpendicular axes (Fig. 4-5). Each possible quantum state of a particle in a box can be represented by a point in this space. For instance, the point α in Fig. 4-5 represents the state $n_1 = 1$, $n_2 = 1$, $n_3 = 1$, and the point β represents the state $n_1 = 2$, $n_2 = 3$, $n_3 = 2$. Note that only positive, nonzero values of n_1, n_2, and n_3 are physically significant, so that "quantum number space" includes only the positive octant of all of (n_1, n_2, n_3) integer space. Furthermore, no quantum states have their points on the n_1, n_2, n_3 axes or on the three coordinate planes. Now Eq. (4-28) is the equation of an ellipsoid with semiaxes N_x, N_y, and N_z, lying along the n_1, n_2, and n_3 axes. The volume of an ellipsoid with semiaxes A, B, C is $(4\pi/3)ABC$. Thus the volume of the octant enclosed by the ellipsoid of Eq. (4-28) and the three coordinate planes is

$$\begin{aligned}\mathfrak{N}(u) &= \frac{1}{8}\cdot\frac{4\pi}{3}\,N_x N_y N_z \\ &= \frac{\pi}{6}\cdot\frac{8m^3 u^3 abc}{h^3} \\ &= \frac{4\pi}{3}\cdot\frac{m^3 V}{h^3}\,u^3\end{aligned} \quad (4\text{-}30)$$

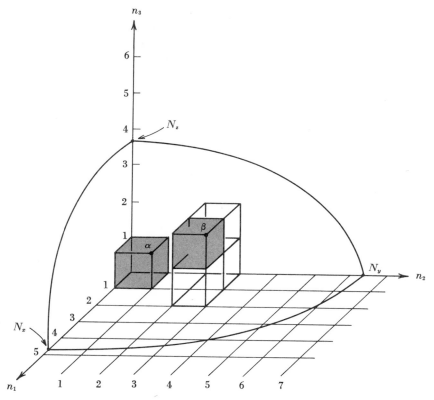

FIG. 4-5 *Quantum state space for a particle in a box. Point α represents the state $n_1 = 1$, $n_2 = 1$, $n_3 = 1$ and point β represents the state $n_1 = 2$, $n_2 = 3$, $n_3 = 2$. To each quantum state can be allocated a cube of volume unity; the cubes associated with states α and β are shaded.*

where we have written $V = abc$, the volume of the box in which the particle moves. The student will readily see from Fig. 4-5 that it is possible to associate with each quantum state a cube of volume unity, and the entire octant can be filled with these cubes. Therefore the number of quantum states lying within the ellipsoid is approximately equal to the volume of the octant enclosed by the ellipsoid. There will be a slight overestimate because some quantum states that have points lying outside the ellipsoid will have portions of their associated cubes lying within the ellipsoid; as the ellipsoids increase in size (N_x, N_y, and N_z increasing), however, the volume of these cubes intersecting the surface of the ellipsoid increases less rapidly than does the number of quantum states completely contained in the ellipsoid. We shall show below that we are ordinarily concerned with values of N_x, N_y, and N_z which are very large (of the order of thousands or millions or more). Therefore we shall make only a small error if we equate $\mathfrak{N}(u)$ to the number of quantum states lying within the ellipsoid—that is, to the number of states whose velocity is less than u.

We may now calculate $S(u, \Delta u)$, the number of quantum states whose velocities lie in the range u to $u + \Delta u$. This number is clearly the number of states whose points in quantum state space lie in the shell between the ellipsoid obtained by setting $u = u$ and the ellipsoid obtained by setting $u = u + \Delta u$ in Eq. (4-28). Thus

$$S(u, \Delta u) = \mathfrak{N}(u + \Delta u) - \mathfrak{N}(u) \qquad (4\text{-}31)$$

Since Δu is assumed to be small, we can write

$$\mathfrak{N}(u + \Delta u) - \mathfrak{N}(u) = \frac{d\mathfrak{N}(u)}{du} \Delta u \qquad (4\text{-}32)$$

so that

$$S(u, \Delta u) = \frac{4\pi m^3 V}{h^3} u^2 \Delta u \qquad (4\text{-}33)$$

Substituting in Eq. (4-27) we find

$$P(u) \Delta u = \frac{1}{Q} \frac{4\pi m^3 V}{h^3} u^2 \exp\left(-\frac{mu^2}{2kT}\right) \qquad (4\text{-}34)$$

or

$$P(u) = Au^2 \exp\left(-\frac{mu^2}{2kT}\right) \qquad (4\text{-}35)$$

where

$$A = \left(\frac{2m^3}{\pi k^3 T^3}\right)^{1/2} \qquad (4\text{-}36)$$

Note that the factor u^2 in the Maxwell-Boltzmann distribution function arises because, as u increases, the number of states, $S(u, \Delta u)$, in the range u to $u + \Delta u$ increases in proportion to u^2, i.e., in proportion to the surface of the ellipsoid in quantum state space. The factor $\exp(-mu^2/2kT)$ in the Maxwell-Boltzmann distribution function occurs because the probability of finding a particle in a single state is proportional to $\exp(-mu^2/2kT)$. Thus

$$\begin{pmatrix} \text{Probability of} \\ \text{finding mole-} \\ \text{cule in veloc-} \\ \text{ity range } u \text{ to} \\ u + \Delta u \end{pmatrix} \propto \begin{pmatrix} \text{Number of} \\ \text{states in} \\ \text{this range} \end{pmatrix} \begin{pmatrix} \text{Probability of} \\ \text{a single state} \\ \text{in this range} \end{pmatrix} \quad (4\text{-}37)$$

or

$$P(u)\,\Delta u \propto (u^2\,\Delta u)\exp\left(-\frac{mu^2}{2kT}\right) \quad (4\text{-}38)$$

The constant A in Eq. (4-35) may also be evaluated by the requirement that

$$\int_0^\infty P(u)\,du = 1 \quad (4\text{-}39)$$

so that

$$A = \left[\int_0^\infty u^2 \exp\left(-\frac{mu^2}{2kT}\right) du\right]^{-1} \quad (4\text{-}40)$$

Setting $x^2 = mu^2/2kT$ we obtain

$$A = \left(\frac{m}{2kT}\right)^{3/2}\left[\int_0^\infty x^2 \exp(-x^2)\,dx\right]^{-1} \quad (4\text{-}41)$$

The integral in Eq. (4-41) has the value $\sqrt{\pi}/4$, so that

$$A = \frac{4}{\sqrt{\pi}}\left(\frac{m}{2kT}\right)^{3/2} \quad (4\text{-}42)$$

This is the same value as that given by Eq. (4-36).

It is worth considering the values of N_x, N_y, and N_z likely to be encountered in a typical gas. If a gas of molecular weight M (molecular mass $m = M/6 \times 10^{23}$ g) is contained in a cubic box with 1 cm sides, then $N_x = N_y = N_z = 2uM/6 \times 10^{23} \times 6.6 \times 10^{-27} \cong 500\, Mu$. Even for hydrogen gas ($M = 2$) and molecules moving at a velocity of only 1 cm/sec (very slow for the molecules in hydrogen gas) we have $N_x \cong 10^3$, so that $\mathfrak{N}(u) \cong 10^9$.

The surface area of the corresponding ellipsoid in quantum state space is only of the order 10^6, so that in equating the volume of the octant of the ellipsoid to the number of quantum states having velocities less than 1 cm/sec we have made an error of only about 0.1%. For heavier gases moving at more nearly the mean thermal velocity, the error will be very much less than this.

In Chapter 5 it will be necessary to know the probability that a gas molecule moves in a particular direction in space, as well as in a specified range of velocities. This probability is easily determined by answering the question: what is the probability that a molecule moves with the x-component of its velocity in the range u_x to $u_x + \Delta u_x$, the y-component of its velocity in the range u_y to $u_y + \Delta u_y$ and the z-component of its velocity in the range u_z to $u_z + \Delta u_z$? Referring to Fig. 4-6 we see that this question has to

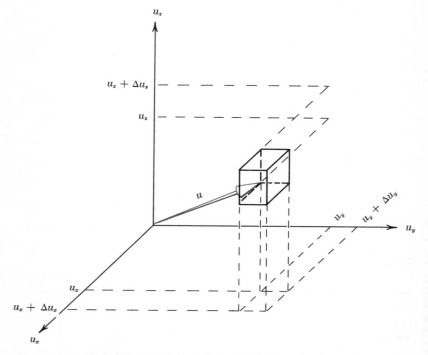

FIG. 4-6 *Quantities involved in the calculation of the probability distribution function for molecular motion in a particular direction.*

do with the probability that the velocity vector **u** lies within the rectangular parallelepiped of volume $\Delta u_x \, \Delta u_y \, \Delta u_z$. This probability is given by the product of $s(u_x, u_y, u_z, \Delta u_x, \Delta u_y, \Delta u_z)$, the number of quantum states corresponding to the range, and

$$(1/Q) \exp\left(-\frac{m(u_x^2 + u_y^2 + u_z^2)}{2kT}\right)$$

the probability of one of these states. But the velocity ranges Δu_x, Δu_y and Δu_z in (u_x, u_y, u_z)-space correspond to the ranges of values Δn_1, Δn_2 and Δn_3 of the translational quantum numbers, n_1, n_2 and n_3, these ranges being related to each other through Eq. (4-20)

$$\Delta u_x = \frac{h}{2ma} \Delta n_1 \qquad \Delta u_y = \frac{h}{2mb} \Delta n_2 \qquad \Delta u_z = \frac{h}{2mc} \Delta n_3 \tag{4-43}$$

The quantity $s(u_x, u_y, u_z, \Delta u_x, \Delta u_y, \Delta u_z)$ is thus the number of states whose quantum numbers lie in the range n_1 to $n_1 + \Delta n_1$, n_2 to $n_2 + \Delta n_2$ and n_3 to $n_3 + \Delta n_3$. This number is the volume in quantum state space corresponding to this range, or

$$s(u_x, u_y, u_z, \Delta u_x, \Delta u_y, \Delta u_z) = \Delta n_1 \, \Delta n_2 \, \Delta n_3 \tag{4-44}$$

Because of Eq. (4-43),

$$\Delta n_1 \, \Delta n_2 \, \Delta n_3 \propto \Delta u_x \, \Delta u_y \, \Delta u_z \tag{4-45}$$

so that the probability we seek is

$$P(u_x, u_y, u_z) \, \Delta u_x \, \Delta u_y \, \Delta u_z$$
$$= A' \, \Delta u_x \, \Delta u_y \, \Delta u_z \exp\left(-\frac{m(u_x^2 + u_y^2 + u_z^2)}{2kT}\right) \tag{4-46}$$

where A' is independent of both the velocity components, u_x, u_y, and u_z, and the ranges Δu_x, Δu_y, and Δu_z. The numerical value of A' is determined by the condition that the total probability of all components of velocity is unity, so that

$$A' \iiint\limits_{\substack{\text{all possible} \\ \text{values of} \\ u_x, u_y, \\ \text{and } u_z}} \exp\left(-\frac{m(u_x^2 + u_y^2 + u_z^2)}{2kT}\right) du_x \, du_y \, du_z = 1 \tag{4-47}$$

or since u_x, u_y, and u_z can each vary independently from $-\infty$ to $+\infty$,

$$A' \int_{-\infty}^{\infty} \exp\left(-\frac{mu_x^2}{2kT}\right) du_x \int_{-\infty}^{\infty} \exp\left(-\frac{mu_y^2}{2kT}\right) du_y$$
$$\int_{-\infty}^{\infty} \exp\left(-\frac{mu_z^2}{2kT}\right) du_z = 1 \quad (4\text{-}48)$$

We introduce the new variables of the type $X = u_x \sqrt{m/2kT}$, giving

$$1 = A' \left(\frac{2kT}{m}\right)^{3/2} \left[\int_{-\infty}^{\infty} \exp(-X^2)\, dx\right]^3 = A' \left(\frac{2\pi kT}{m}\right)^{3/2} \quad (4\text{-}49)$$

since $\int_{-\infty}^{\infty} \exp(-X^2)\, dX = \sqrt{\pi}$. Thus the probability in question is

$$P(u_x, u_y, u_z)\, \Delta u_x\, \Delta u_y\, \Delta u_z$$
$$= \left(\frac{m}{2\pi kT}\right)^{3/2} \exp\left(-\frac{m(u_x^2 + u_y^2 + u_z^2)}{2kT}\right) \Delta u_x\, \Delta u_y\, \Delta u_z \quad (4\text{-}50)$$

EXERCISE Show that about 10^{10} quantum states are available to a helium atom in a cubical box 1 cm on a side, moving with u_x, u_y, and u_z each between 1000 cm/sec and 1001 cm/sec.

EXERCISE If one transforms from cartesian coordinates to polar coordinates in Eq. (4-50) one may write $\Delta u_x\, \Delta u_y\, \Delta u_z = u^2 \sin\theta\, \Delta u\, \Delta\theta\, \Delta\phi$. Show that if one introduces this into Eq. (4-50) and integrates over θ and ϕ, one obtains the Maxwell-Boltzmann distribution, Eqs. (4-35) and (4-36).

4-4 Some properties of the Maxwell-Boltzmann distribution

In general, if one has an unnormalized distribution function $P(x)$ in some variable x, and if some property $g(x)$ is associated with x, then the mean value of the property g is defined by

$$\bar{g} = \int_{\text{all } x} g(x) P(x)\, dx \bigg/ \int_{\text{all } x} P(x)\, dx \quad (4\text{-}51)$$

This definition (which is known as the *mean value principle* and has already been used in Section lb of Chapter 2) is consistent with our everyday notion of the average values of properties. For

DISTRIBUTION OF MOLECULAR VELOCITIES

instance, let $x = 1$ stand for apples, $x = 2$ for oranges and $x = 3$ for peaches. If there are three apples, four oranges and two peaches we can write $P(1) = 3$, $P(2) = 4$, $P(3) = 2$; if apples cost 5 cents, oranges 10 cents, and peaches 8 cents apiece, we can write $g(1) = 5¢$, $g(2) = 10¢$, $g(3) = 8¢$. The average price of a piece of fruit is then

$$\bar{g} = \frac{g(1)P(1) + g(2)P(2) + g(3)P(3)}{P(1) + P(2) + P(3)} = \frac{\sum_{\text{all } x} g(x)P(x)}{\sum_{\text{all } x} P(x)}$$

$$= \frac{5¢ \times 3 + 10¢ \times 4 + 8¢ \times 2}{3 + 4 + 2} = 7\tfrac{8}{9}¢ \qquad (4\text{-}52)$$

Equation (4-51) is an obvious extension of Eq. (4-52) for the case in which x can vary continuously.

Let us apply Eq. (4-51) to the calculation of the average values of various properties of gas molecules which depend on their velocity, u. First let us calculate the mean value of the velocity itself. Setting $x = u$, $g(x) = u$ and $P(x) = Au^2 \exp(-mu^2/2kT)$ in Eq. (4-51) we obtain

$$\bar{u} = \int_0^\infty u \cdot Au^2 \exp\left(-\frac{mu^2}{2kT}\right) du \bigg/ \int_0^\infty Au^2 \exp\left(-\frac{mu^2}{2kT}\right) du$$

$$= \int_0^\infty u^3 \exp\left(-\frac{mu^2}{2kT}\right) du \bigg/ \int_0^\infty u^2 \exp\left(-\frac{mu^2}{2kT} du\right) \qquad (4\text{-}53)$$

The definite integrals may be evaluated using the formulas

$$\int_0^\infty x^{2n+1} \exp(-ax^2)\, dx = \frac{n!}{2a^{n+1}} \qquad (4\text{-}54)$$

$$\int_0^\infty x^{2n} \exp(-ax^2)\, dx = \frac{1 \cdot 3 \cdot 5 \cdots (2n-1)}{2^{n+1}} \sqrt{\frac{\pi}{a^{2n+1}}} \qquad (4\text{-}55)$$

where Eq. (4-54) is to be used for integrals in which the variable factor ahead of the exponential is raised to an odd power [as in the numerator of Eq. (4-53)] and Eq. (4-55) is to be used if this variable is raised to an even power [as in the denominator of Eq. (4-53)].

Substitution of Eqs. (4-54) and (4-55) in Eq. (4-53), with $a = m/2kT$ and $n = 1$ in both integrals, gives

$$\bar{u} = \frac{1!/2(m/2kT)^2}{(1/2^2)\sqrt{\pi/(m/2kT)^3}} = \sqrt{\frac{8kT}{\pi m}} = \sqrt{\frac{8RT}{\pi M}} \qquad (4\text{-}56)$$

where R is the gas constant and M is the molecular weight of the gas.

To calculate the mean-square velocity, $\overline{u^2}$, we set $g(x) = u^2$ in Eq. (4-51) and find

$$\overline{u^2} = \int_0^\infty u^4 \exp\left(-\frac{mu^2}{2kT}\right) du \bigg/ \int_0^\infty u^2 \exp\left(-\frac{mu^2}{2kT}\right) du$$

$$= \frac{1\cdot 3}{2^3} \sqrt{\frac{\pi}{(m/2kT)^5}} \bigg/ \frac{1}{2^2} \sqrt{\frac{\pi}{(m/2kT)^3}}$$

$$= \frac{3kT}{m} = \frac{3RT}{M} \qquad (4\text{-}57)$$

which agrees with the result obtained in Chapter 2 [Eq. (2-38)].

To calculate the mean kinetic energy, $\bar{\epsilon}$, we set $g(x) = \frac{1}{2}mu^2$ and obtain

$$\bar{\epsilon} = \frac{1}{2} m \overline{u^2} = \frac{3}{2} kT \qquad (4\text{-}58)$$

in agreement with the result of Chapter 3 [Eq. (3-7)].

The most probable velocity, u_{mp} is the velocity that gives $P(u)$ its maximum value. This is the value of u that makes $dP/du = 0$,

$$\frac{dP(u)}{du} = \frac{d(Au^2 \exp(-mu^2/2kT))}{du}$$

$$= A\left[2u \exp\left(-\frac{mu^2}{2kT}\right) + u^2\left(-\frac{mu}{kT}\right)\exp\left(-\frac{mu^2}{2kT}\right)\right]$$

$$= Au \exp\left(-\frac{mu^2}{2kT}\right)\left[2 - \frac{mu^2}{kT}\right] \qquad (4\text{-}59)$$

which vanishes when $u = 0$, $\sqrt{2kT/m}$ and ∞; the root we seek is

$$u_{\text{mp}} = \sqrt{2kT/m} = \sqrt{2RT/M} \qquad (4\text{-}60)$$

Consider next the quantity

$$F(u) = \int_u^\infty P(u)\, du \qquad (4\text{-}61)$$

which is the fraction of molecules whose velocity is greater than u. The value of u for which $F(u)$ is one-half is known as the median velocity, u_{med}

$$F(u_{\text{med}}) = 0.50 \tag{4-62}$$

It is found (see exercise below) that

$$\begin{aligned} u_{\text{med}} &= 1.538 \sqrt{kT/m} \\ &= 1.09 u_{\text{mp}} \end{aligned} \tag{4-63}$$

The four velocities, $u_{\text{mp}}, u_{\text{med}}, \bar{u}$, and u_{rms} [the root-mean-square velocity, $(\overline{u^2})^{1/2}$] are in the ratio

$$\begin{aligned} u_{\text{mp}} : u_{\text{med}} : \bar{u} : u_{\text{rms}} &= \sqrt{2} : 1.538 : \sqrt{8/\pi} : \sqrt{3} \\ &= 1 : 1.09 : 1.12 : 1.225 \end{aligned} \tag{4-64}$$

These four velocities are thus similar in magnitude but do not have exactly the same values. Figure 4-3 shows these velocities in relation to the distribution function.

If the temperature in the Maxwell-Boltzmann distribution function is replaced by the expression

$$T = \frac{m u_{\text{mp}}^2}{2k} \tag{4-65}$$

we have

$$\begin{aligned} P(u) &= A u^2 \exp\left(-\frac{u^2}{u_{\text{mp}}^2}\right) du \\ &= (4/\sqrt{\pi})(u/u_{\text{mp}})^2 \exp\left(-\frac{u^2}{u_{\text{mp}}^2}\right) d(u/u_{\text{mp}}) \end{aligned} \tag{4-66}$$

Writing

$$x = \frac{u}{u_{\text{mp}}} \tag{4-67}$$

then

$$P(u)\, du = \frac{4}{\sqrt{\pi}} x^2 \exp(-x^2)\, dx \tag{4-68}$$

and

$$F(u) = \frac{4}{\sqrt{\pi}} \int_{u/u_{\text{mp}}}^{\infty} x^2 \exp(-x^2)\, dx \tag{4-69}$$

TABLE 4-1 VALUES OF $F(u)$ [a] FOR VARIOUS VALUES OF u/u_{mp} [b]

u/u_{mp}	0	0.354	0.707	1.061	1.087	1.414	1.768	2.122	2.476
$F(u)$	1.000	0.969	0.801	0.522	0.500	0.261	0.100	0.030	0.0066

u/u_{mp}	3	5	7	10	15	20
$F(u)$	4.5×10^{-4}	4×10^{-10}	4×10^{-21}	4×10^{-43}	5×10^{-97}	6×10^{-173}

[a] The fraction of molecules having a velocity greater than u.
[b] u_{mp} is the most probable velocity ($= \sqrt{2RT/M}$).

Table 4-1 shows values of $F(u)$ for various values of u/u_{mp}. It is evident that $F(u)$ drops off very quickly as u exceeds the most probable velocity. Only about 3% of the molecules have velocities in excess of double the most probable velocity, and only 0.05% have velocities greater than $3u_{mp}$. In one mole of gas, only about 100 molecules move with velocities greater than $7u_{mp}$ and in order to be reasonably confident that one had a single molecule with a velocity $10u_{mp}$ one would need to have about 10^{20} moles of gas.

EXERCISE Show that

$$F(x) = \frac{2}{\sqrt{\pi}} \left[x \exp(-x^2) + \int_x^\infty \exp(-x^2)\, dx \right]$$
$$= 1 - 2I(\sqrt{2}\, x) + 2xG(\sqrt{2}\, x)$$

where $G(y)$ is the Gauss error function,

$$G(y) = \frac{1}{\sqrt{2\pi}} \exp\left(-\frac{y^2}{2}\right)$$

and $I(y)$ is the Gauss error integral

$$I(y) = \frac{1}{\sqrt{2\pi}} \int_0^y \exp\left(-\frac{y^2}{2}\right) dy$$

values of which are tabulated in handbooks. Show that if $x \gg 1$, then it is a good approximation to write

$$\int_x^\infty \exp(-x^2)\, dx \cong \frac{1}{2x} \exp(-x^2)$$

[*Hint:* replace the variable x in the integrand by $x_0 + y$ where y is a new variable and x_0 is the lower limit of the error integral. Then $\exp(-x^2) =$

$\exp(-y^2 - 2x_0 y - x_0^2)$ and

$$\int_x^\infty \exp(-x^2)\, dx = \exp(-x_0^2) \cdot \int_0^\infty \exp(-y^2) \exp(-2x_0 y)\, dy$$

If $x_0 \gg 1$, the factor $\exp(-y^2)$ will have the value unity for values of y which give nonnegligible values of $\exp(-2x_0 y)$, so we may write

$$\int_x^\infty \exp(-x^2)\, dx \cong \exp(-x_0^2) \int_0^\infty \exp(-2x_0 y)\, dy]$$

Thus if $x \gg 1$

$$F(x) \cong \frac{\exp(-x^2)}{\sqrt{\pi}} \left(2x + \frac{1}{x}\right)$$

Use this formula and the exact expression to check some of the values given in Table 4-1.

The effect of temperature on the Maxwell-Boltzmann distribution function is shown in Fig. 4-7. As the temperature is raised

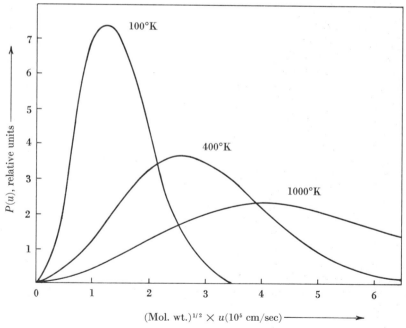

FIG. 4-7 *Effect of temperature on the Maxwell-Boltzmann distribution.*

TABLE 4-2 EFFECTS OF TEMPERATURE ON $F(u)$[a] FOR VARIOUS VALUES OF u

Temp. (°K)	u_{mp} (cm/sec)	$F(u)$ $u = 2.5 \times 10^4$ cm/sec	$u = 5 \times 10^4$ cm/sec	$u = 10 \times 10^4$ cm/sec	$u = 20 \times 10^4$ cm/sec	$u = 30 \times 10^4$ cm/sec
100	2.28×10^4	0.384	0.018	2.4×10^{-8}	5×10^{-33}	1×10^{-74}
200	3.21×10^4	0.607	0.137	2.3×10^{-4}	1×10^{-16}	2×10^{-37}
300	3.94×10^4	0.708	0.272	5×10^{-3}	1×10^{-11}	3×10^{-34}
400	4.56×10^4	0.771	0.384	1.6×10^{-2}	2×10^{-8}	2×10^{-18}

[a] The fraction of oxygen molecules moving with velocities greater than u

there is a decrease in the fraction of the slower moving molecules and an increase in the fraction of the more rapidly moving ones. The distribution becomes broader and the peak of the distribution moves to higher velocities. Perhaps the most striking effect of the temperature is on those relatively few molecules which move with very high velocity. This effect is seen in Table 4-2, where $F(u)$ is shown for oxygen for various values of u and for various temperatures. Raising the temperature from 300°K to 400°K increases $F(u)$ by about 10% if $u = 25,000$ cm/sec ($\cong 0.6 u_{mp}$ at these temperatures), by about 40% if $u = 50,000$ cm/sec ($\cong 1.2 u_{mp}$) and by a factor of 3.2 if $u = 100,000$ cm/sec ($\cong 2.5 u_{mp}$ at these temperatures); but if $u = 200,000$ cm/sec ($\cong 5.0 u_{mp}$), this factor is 2000, and if $u = 300,000$ cm/sec ($\cong 7.5 u_{mp}$), the factor is nearly 10^{16}. This effect is responsible for the large temperature coefficient typically observed for the rates of chemical reactions. In many chemical reactions, rearrangements of the valence bonds between atoms are possible only if the molecules involved in the reaction collide with a sufficiently violent impact—that is, at least one of the colliding molecules must move with sufficiently high velocity to provide the activation energy required for the reactants to undergo chemical change. This velocity is typically of the order of $5 u_{mp}$ for a reaction that goes at a measureable rate. The number of these high velocity molecules increases very rapidly as their temperature is raised.

EXERCISE It is frequently observed that the rates of chemical reactions increase by a factor of between two and three when the temperature is raised by 10°C at room temperature. Show that this indicates that molecules moving with at least 5 to 6 times the most probable velocity determine the rate of the reaction. [*Hint:* Setting $x = u/u_{mp}$ show that for a given value of u, $dx/dT = -x/2T$ and that $d \ln F/dx \cong -2x$ if $x \gg 1$. Set $d \ln F = \ln 2$ or $\ln 3$ and $dT = 10°$, $T = 300°K$.]

Problems

1. (a) Show that the mean velocity of a nitrogen molecule at 300°K is 476 m/sec. (b) What is the ratio of the probability of finding a nitrogen molecule whose velocity is 476 m/sec at 300°K to the probability of finding a nitrogen molecule with the same velocity at 310°K? (c) What is the corresponding ratio for a nitrogen molecule with a speed in the vicinity of $3 \times 476 = 1428$ m/sec?

2. Prove that in a gas with a Maxwell-Boltzmann distribution of molecular velocities the fraction of molecules whose kinetic energy, $\epsilon = \frac{1}{2}mu^2$, lies between ϵ and $\epsilon + d\epsilon$ is proportional to

$$\epsilon^{1/2} \exp\left(-\frac{\epsilon}{kT}\right) d\epsilon$$

3. A computer has been programmed to produce the numerical value of the integral

$$I_n(y) = \int_y^\infty x^n \exp(-x^2)\, dx$$

for any values of y and n that the computer operator indicates. Show how you would use this information to find (a) the fraction of argon atoms whose velocities are greater than 1000 m/sec at 300°K; (b) the mean velocity of these fast molecules; (c) the mean energy of these fast molecules.

4. In order for any object to escape from the earth's gravitational field it must have a velocity greater than 11 km/sec. What fraction of the nitrogen molecules of the air at 300°K have at least this "escape velocity?" What fraction of helium atoms at 300°K have a velocity greater than the escape velocity? If the earth's surface temperature were suddenly raised to 1300°K, what fraction of the nitrogen molecules in the air would have velocities greater than the escape velocity?

5. Find $\overline{u^3}$ for a gas of molecular weight M at a temperature T.

6. (a) An electron (mass $\frac{1}{1836}$ that of a proton) moves in a cubical box 10^{-8} cm on a side. How many quantum states are there with energy less than kT at 300°K? (b) How many quantum states would there be with energy less than kT if the electron moved in a cubical box 1 cm on a side? (c) The density of metallic sodium is 0.97 g/cm³ and

its atomic weight is 23. Each atom has a single valence electron. How many valence electrons are there in 1 cm³ of metallic sodium? (d) The valence electrons of metallic sodium are believed to move about more or less as if they were a gas. Because of the Pauli principle and because of the two spin states of an electron, two electrons can be placed in each translational energy state. How does the number of electrons per cm³ in sodium metal compare with the number of quantum states available to electrons with energy less than kT at 300°K? (e) At what temperature would the number of quantum states available to electrons with energy less than kT be equal to the number of valence electrons in 1 cm³ of sodium metal?

7. A narrow, well-defined beam of molecules is produced by allowing a gas at temperature T to effuse through a small hole from a chamber containing the gas at a finite pressure into an evacuated space. A system of collimating slits is lined up in the evacuated space, removing all molecules whose direction of motion does not lie within a small solid angle of the axis of the collimating system. Show that the mean x-component of velocity of the molecules in the beam is $\bar{u}_x = (2kT/\pi m)^{1/2}$ and the mean square x-component of velocity is $\overline{u_x^2} = kT/m$, where m is the molecular mass. [*Hint:* recall that the distribution function for velocities in a specific direction, x, is $P(u_x) \propto \exp(-mu_x^2/2kT)$.]

8. A spherical pot of perfumed sugar water hangs from a tree. The radius of the pot is a. After some time insects are attracted to the pot and it is found that the average number of insects per unit volume at a distance r from the center of the pot is given by K/r^n, where K and n are empirical constants and where $r > a$. The total number of insects flying around the pot is N. (a) Show that K, N, a, and n are related by $K = (n-3)Na^{n-3}/4\pi$. (b) What is the minimum possible value of n if the above relation is to hold at all values of r from $r = a$ to $r = \infty$? (c) Show that the probability of finding a particular insect in the region between r and $r + dr$ is $(n-3)(a/r)^{n-2}(dr/a)$. (d) Find $\bar{r}$, the mean distance of an insect from the pot. (e) Find r_{mp}, the most probable distance between an insect and the pot.

9. From Eq. (4-30) we find that the number $\mathfrak{N}(u)$ of quantum states available to a particle with velocity less than u moving in a three dimensional box is proportional to u^3. Since the energy ϵ is proportional to u^2, the number of quantum states with energy less than ϵ is $\mathfrak{N}(\epsilon) \propto \epsilon^{3/2}$. Show that for a particle moving in a two-dimensional box, $\mathfrak{N}(\epsilon) \propto \epsilon$, and for a particle moving in a one-dimensional box, $\mathfrak{N}(\epsilon) \propto \epsilon^{1/2}$.

10. Two identical particles, each of mass m, move in a cubic box whose sides are of length a. The energy levels of the system are

$$\epsilon = (h^2/8ma^2)[n_{1x}^2 + n_{1y}^2 + n_{1z}^2 + n_{2x}^2 + n_{2y}^2 + n_{2z}^2]$$

where n_{it} is the quantum number for particle i ($i = 1$ or 2) moving in direction t ($t = x, y,$ or z). If we write $A^2 = 8ma\epsilon^2/h^2$, this equation can be written in the form

$$A^2 = n_{1x}^2 + n_{1y}^2 + n_{1z}^2 + n_{2x}^2 + n_{2y}^2 + n_{2z}^2$$

This is the equation for a hypersphere of radius A in an imaginary six-dimensional space defined by six mutually perpendicular axes, n_{1x}, n_{1y}, n_{1z}, n_{2x}, n_{2y}, and n_{2z}. It can be shown that the volume enclosed by a hypersphere in m dimensions is proportional to A^m, where A is the radius. This volume is also proportional to the number of quantum states having energies less than ϵ. Thus for two particles, each moving in three dimensions, the number of states available with energy less than ϵ is proportional to A^6, which is in turn proportional to ϵ^3. By similar arguments we see that if N particles move in a three dimensional box the number, $\mathfrak{N}(\epsilon, N)$, of states with energy less than ϵ is proportional to $\epsilon^{3N/2}$. Furthermore, the number of states with energies between ϵ and $\epsilon + \Delta\epsilon$ is

$$\Delta\mathfrak{N} = \frac{d\mathfrak{N}}{d\epsilon}\Delta\epsilon \propto \epsilon^{(3N/2)-1}\Delta\epsilon$$

Thus if a monatomic gas with N atoms is given a total translational energy in a fixed interval ΔE in the vicinity of energy E, the number of quantum states available to it is proportional to $E^{(3N/2)-1}$. Let one of the molecules be placed in a state having an energy ϵ, the remaining energy of the system, $E - \epsilon$, being distributed at random among the remaining $N - 1$ molecules. If the mean translational energy per molecule is defined as $\bar{\epsilon} = E/N$, show that $P(\epsilon)$, the probability of finding a molecule in a state with energy ϵ, is given by

$$P(\epsilon) \propto \exp\left(-\frac{3\epsilon}{2\bar{\epsilon}}\right)$$

If the mean translational energy is identified with $\frac{3}{2}kT$ [as shown in Chapter 3, Section 1, Eq. (4-7)], we obtain

$$P(\epsilon) \propto \exp\left(-\frac{\epsilon}{kT}\right)$$

which is the Boltzmann factor. [*Hint:* refer to Appendix 3-1.]

Supplementary references

Elementary derivations of the Maxwell-Boltzmann distribution from a classical point of view will be found in:

G. W. Castellan, *Physical Chemistry*, Addison-Wesley, Reading, Mass., 1964, pp. 52–57.

W. J. Moore, *Physical Chemistry*, 3rd ed., Prentice-Hall, Englewood, N. J., 1962, pp. 232–238.

The experimental verification of the Maxwell-Boltzmann distribution is described in Jeans' *Introduction* (see References, Chapter 2), pp. 124–130. See also H. S. Taylor and S. Glasstone, *Treatise on Physical Chemistry*, D. Van Nostrand, New York, 1951, Vol. II, pp. 35–40.

Chapter 5

MOLECULAR COLLISIONS AND THE TRANSPORT PROPERTIES OF GASES

IN ORDER for two molecules to undergo a chemical reaction, such as $H_2 + I_2 \rightarrow 2HI$, it is necessary for them to interact with each other—that is, to collide. Furthermore, the probability that a given pair of molecules will react in a collision can be expected to depend very strongly on the detailed dynamics of the collision— the relative velocities of the centers of gravity of the molecules (how violently they hit one another), how close they come to each other (whether the collision is grazing or head-on), the relative orientations of the molecules in space, whether or not the molecules are rotating or vibrating, and so forth. The detailed understanding of molecular collisions is therefore of considerable interest to chemists. Indeed, the investigation of collision dynamics is for this reason one of the most active research areas in physical chemistry today. In addition, a number of important physical properties of gases (viscosity, heat conductivity and diffusion) depend in an important way on the dynamics of molecular collisions. These properties are interesting both intrinsically and because of the insight into chemical collisions that can, at least potentially, be derived from understanding them.

In this chapter we shall study the dynamics of collisions and some of the physical properties which depend on molecular collisions. Two important approaches are available for dealing with collisions in gases. In one of these approaches molecules are assumed to be rigid objects with definite boundaries and exerting no attractive forces on each other. A collision between a pair of these objects is a well-defined event (as it is between a pair of billiard balls). If a gas consisted of a large number of such molecules, one could speak of collision frequencies and of the average distance that a molecule travels between successive collisions. This molecular model has the advantage that it leads to a relatively simple theory of many gas properties which depend on molecular collisions. It can, however, give only approximately correct results because molecules are really not rigid objects with sharply defined interaction boundaries. Since the interaction potential between a pair of molecules is finite at all distances—though perhaps very small at large distances—it is impossible to establish a clear definition of a molecular collision. In a sense, every molecule in a gas or liquid is colliding (or interacting) at all times with all other molecules in the sample. To avoid this difficulty, a second kind of approach has been developed which is based on an examination of the precise nature of molecular collisions in the light of all that we know about intermolecular forces. This approach is, however, very much more difficult to understand. In this chapter we shall devote most of our attention to the first approach and be satisfied with a quotation of results from the second approach.

5-1 *Collision frequency in a gas*

a. COLLISION FREQUENCY BETWEEN LIGHT, RAPIDLY MOVING MOLECULES AND HEAVY, SLOWLY MOVING ONES Consider a gas mixture containing two components, one of low molecular weight, and hence high mean molecular velocity, the other of high molecular weight and low mean molecular velocity. This might correspond to a mixture of hydrogen (molecular weight 2) and xenon (atomic weight 131), in which the hydrogen molecules move on the average about 8 [$\cong (131/2)^{1/2}$] times as rapidly as xenon atoms. It could also apply to a mixture of electrons (molecular weight 1/1840) and ordinary molecules (say air, molecular weight 29) in which the electrons move about 230 [$\cong (1840 \times 29)^{1/2}$] times as fast as the air molecules. The atoms, molecules and electrons will be assumed to be rigid spheres of definite size, exerting no attractive forces on one another. How many collisions take place in one second between a given light molecule and the heavy ones?

Let $\sigma = \frac{1}{2}(\sigma_1 + \sigma_2)$ be the average of the diameters of the light molecule (σ_1) and of a heavy molecule (σ_2); σ is the distance between the centers of the two molecules at the instant of collision—the *collision diameter* (Fig. 5-1a). We can imagine that a light molecule, as it moves through space, sweeps out a cylinder of radius σ and base area $\pi\sigma^2$. It is evident that any heavy molecule whose center lies within this cylinder will undergo a collision with the light molecule (Fig. 5-1b, in which light molecule a will suffer no collision with heavy molecule A, a grazing collision with B and a severe collision with C). If in one second the light molecule moves a distance U, then the volume swept out in one second by the light molecule is $\pi\sigma^2 U$. If there are n heavy molecules in unit volume, and if the motion of the heavy molecules can be neglected, then the number of collisions per second between a given light molecule and the heavy molecules is

$$z = \pi\sigma^2 U n \qquad (5\text{-}1)$$

If there are N light molecules per unit volume, then the total number of collisions between light and heavy molecules per second and per unit volume is

$$Z = \pi\sigma^2 U n N \qquad (5\text{-}2)$$

These formulas are correct only if the distance traveled between collisions is large compared with the collision diameter. Because collisions change the direction of motion, the path of a molecule is not a straight line. Consequently the volume swept out by a molecule is a zig-zag cylinder (Fig. 5-1c), bent at each collision. Thus in writing $\pi\sigma^2 U$ for the volume swept out in unit time we assume that the modifications caused by the bends can be neglected. We shall show later that, at pressures below a few atmospheres and at ordinary temperatures, the distance between colli-

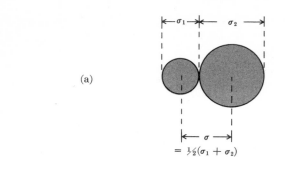
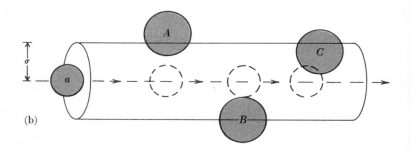
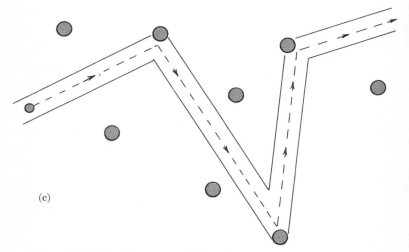

FIG. 5-1 *Collision diameter and mean free path.*

MOLECULAR COLLISIONS

sions is typically very much greater than a molecular diameter. Under these conditions Eqs. (5-1) and (5-2) will give reasonably accurate collision frequencies for gas mixtures of the appropriate type.

The collision diameter of a hydrogen molecule is about 2.5Å and that of a xenon atom is about 4.9Å. One mole of a gas at 1 atm and 300°K occupies about 24,600 cm^3, so for xenon under these conditions $n = (6.023 \times 10^{23}$ molecules/mole)/(24,600 cm^3/mole) $= 2.45 \times 10^{19}$ molecules/cm^3. The mean velocity of H$_2$ is

$$U = (8RT/\pi M)^{1/2} = [8 \times (8.314 \times 10^7 \text{ ergs/deg mole}) \times (300 \text{ deg})/\pi(2 \text{ g/mole})]^{1/2} = 5.75 \times 10^5 \text{ cm/sec}$$

so that the collision rate for a given hydrogen molecule with xenon atoms at 1 atm and 300°K is

$$z = \pi \times \left(\frac{2.5 + 4.9}{2}\right) \times 10^{-16} \text{ cm}^2 \times (5.75 \times 10^5 \text{ cm/sec})$$
$$\times (2.45 \times 10^{19} \text{ molecule/cm}^3)$$
$$= 1.64 \times 10^{10} \text{ collisions/sec}$$

EXERCISE How many collisions per second does a given xenon atom have with hydrogen molecules if the hydrogen pressure is 1 atm and its temperature is 300°K?

b. COLLISION FREQUENCY BETWEEN IDENTICAL GAS MOLECULES MOVING WITH THE SAME SPEED In the situation just considered, it has been assumed that the heavy molecules are not moving, so that the absolute velocity of the light molecules is identical with their velocity relative to the heavy molecules. If both molecules involved in a collision are moving, however, the calculation must be modified. Consider a molecule A of diameter σ, moving with velocity u, on which impinges a beam of identical molecules moving with the same velocity but in a direction making an angle θ with the motion of A (Fig. 5-2a). Let the density of molecules in the beam be ν molecules per unit volume. Then the number of beam molecules whose centers pass, in time dt, within a distance σ of the center of A, and hence collide with A, is the number of beam molecules contained in a cylinder of base $\pi\sigma^2$ and height $u_{\text{rel}} \, dt$, where u_{rel} is the velocity of the beam molecules as seen by an observer traveling with A. Therefore the number of collisions in unit time is

$$z = \nu\pi\sigma^2 u_{\text{rel}} \tag{5-3}$$

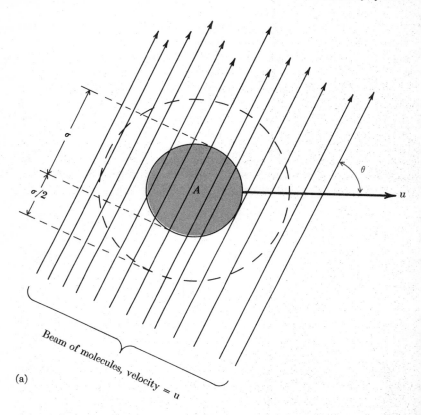

(a)

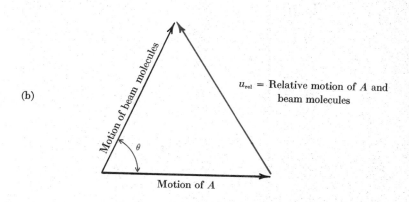

(b)

According to the law of cosines, if a, b, and c are the sides of a triangle and θ is the angle between a and b, then

$$c = (a^2 + b^2 - 2ab \cos \theta)^{1/2} \tag{5-4}$$

Drawing the vector diagram for the velocities of A and of the beam (Fig. 5-2b) we observe that we may set $a = b = u$ in Eq. (5-4) so that

$$u_{\text{rel}} = u[2(1 - \cos \theta)]^{1/2} \tag{5-5}$$

and

$$z = \nu \pi \sigma^2 u[2(1 - \cos \theta)]^{1/2} \tag{5-6}$$

Thus if $\theta = 0°$, so that the beam moves in the same direction as A and $\cos \theta = 1$, then $z = 0$ and there can be no collisions at all. On the other hand, if $\theta = 180°$ ($\cos \theta = -1$), corresponding to the motion of A and of the beam in opposite directions, then $u_{\text{rel}} = 2u$ so that $z = 2\nu\pi\sigma^2 u$. If $\theta = 90°$, then $z = 2^{1/2} \nu\pi\sigma^2 u$ (beam moving at right angles to A). Clearly the relative motion of the colliding molecules has a most important effect on their collision frequency, and it must be taken into account in calculating the collision frequency in, say, a pure gas, or in a gas consisting of molecules whose molecular weights do not differ greatly.

In order to determine from Eq. (5-6) the collision frequency in a pure gas, it is necessary to relate ν, the molecular concentration in the beam, to n, the number of molecules in unit volume of the gas. Consider the increment $d\Omega$ of solid angle defined by the ranges θ, $\theta + d\theta$, ϕ, $\phi + d\phi$, where θ and ϕ are the polar angles shown in Fig. 5-3, and where the angle θ is the same as that appearing in Eqs. (5-4), (5-5), and (5-6) (that is, the z-axis, $\theta = 0$, lies along the direction of motion of molecule A). Since $d\Omega$ is defined as the area on a sphere of unit radius enclosed by the two great circles ϕ and

---

FIG. 5-2 (a) *Geometry of collisions between a molecule, A, moving with a velocity u, and a beam of molecules identical with A and moving with the same velocity as A but in a direction making an angle θ with the direction of motion of A. The diameter, σ, of the full circle drawn around A is the collision diameter; all beam molecules whose centers pass within a distance σ of the center of A will collide with A.* (b) *Vector diagram for finding the relative velocity, u_{rel}, of A and the beam.*

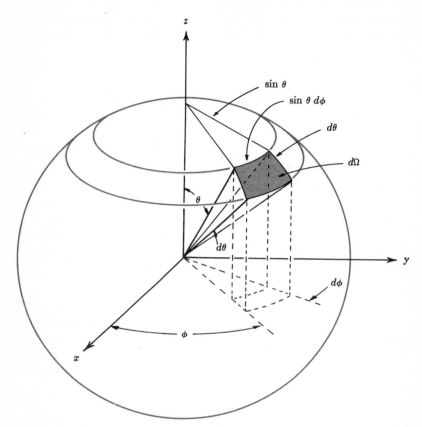

FIG. 5-3 Polar angles used in Eq. (5-7). The angle ϕ is the longitude and θ is the colatitude (i.e., the latitude measured from the North Pole rather than from the Equator). The solid angle, $d\Omega$, is the area on the surface of a unit sphere of the quasirectangular region whose sides are $\sin\theta\, d\phi$ and $d\theta$. The total surface of the unit sphere is 4π, so the fraction of the total surface covered by $d\Omega$ is $d\Omega/4\pi$.

$\phi + d\phi$ and the two arcs θ and $\theta + d\theta$, we have (cf. Fig. 5-3)

$$d\Omega = \sin\theta\, d\theta\, d\phi \tag{5-7}$$

The total surface area of the unit sphere is 4π,

$$\int_{\substack{\text{all} \\ \text{directions}}} d\Omega = \int_0^{2\pi} d\phi \int_0^{\pi} \sin\theta\, d\theta = 4\pi \tag{5-8}$$

MOLECULAR COLLISIONS

Since molecules move in all directions with equal probability, the number of molecules in unit volume, dn, moving in the directions enclosed by the solid angle $d\Omega$ must be

$$dn = n(d\Omega/4\pi) = (n/4\pi) \sin\theta \, d\theta \, d\phi \tag{5-9}$$

(The factor 4π in the denominator of Eq. (5-9) assures that

$$\int_{\substack{\text{all}\\\text{directions}}} dn = n)$$

We shall now assume that all molecules in the gas have the same velocity, u—an assumption which, of course, is not strictly correct, but which will be modified in Section 5-1c, below. Then we may equate the quantity ν in Eq. (5-6) with dn in Eq. (5-9) and obtain dz, the number of collisions in unit time between a given molecule A and all molecules in the gas moving in the solid angle $d\Omega$

$$dz = (\pi\sigma^2 u)\frac{n}{4\pi}[2(1-\cos\theta)]^{1/2}\sin\theta \, d\theta \, d\phi \tag{5-10}$$

Integrating over all directions we obtain for the total collision frequency of a single molecule

$$z = \frac{\sqrt{2}\,\pi\sigma^2 un}{4\pi}\int_0^{2\pi} d\phi \int_0^\pi (1-\cos\theta)^{1/2}\sin\theta \, d\theta$$

$$= -\frac{\pi\sigma^2 un}{\sqrt{2}}\int_0^\pi (1-\cos\theta)^{1/2}\, d\cos\theta$$

$$= \frac{\pi\sigma^2 un}{\sqrt{2}}\int_{-1}^1 (1-x)^{1/2}\, dx$$

Setting $y = 1 - x$ in the integral we obtain

$$\int_{-1}^1 (1-x)^{1/2}\, dx = \int_0^2 y^{1/2}\, dy = \frac{4}{3}\sqrt{2}$$

so that

$$z = \left(\frac{4\pi}{3}\right)\sigma^2 un \tag{5-11}$$

Comparison with Eq. (5-2) reveals that the random relative motion of the molecules increases the collision rate by a factor $\frac{4}{3}$; but it should be emphasized that we have assumed that all molecules have the same velocity, which is, of course, never the case in a real gas.

In order to obtain the total number of collisions Z between all molecules in unit volume and in unit time, the collision frequency of a single molecule must be multiplied by the total number of molecules in unit volume. If unit volume of the gas contains n_A molecules of type A and n_B molecules of type B, then the number of collisions in unit time and in unit volume between A and B molecules will be

$$Z_{AB} = n_A z_{AB} = \frac{4}{3}\pi\sigma^2 u n_A n_B \qquad (5\text{-}12\text{a})$$

where z_{AB} is the collision rate of a particular A molecule with B molecules. On the other hand, in computing the collision frequency, Z_{AA}, between A molecules, the product $n_A z_{AA}$ will count each collision between A molecules twice (z_{AA} being the collision rate of a particular A molecule with other A molecules). Thus

$$Z_{AA} = \frac{1}{2} n_A z_{AA} = \frac{2}{3}\pi\sigma^2 u n_A^2 \qquad (5\text{-}12\text{b})$$

If u in Eqs. (5-11) and (5-12) is replaced by the mean thermal velocity, $\bar{u}$, we obtain an approximate collision number

$$Z_{AA} = \frac{2}{3}\pi\sigma^2 n_A^2 \bar{u} = \frac{2}{3}\pi\sigma^2 n_A^2 (8RT/\pi M)^{1/2} \qquad (5\text{-}13)$$

Equations (5-10), (5-12) and (5-13) are approximate because they are based on the assumption that all molecules move with the same speed. We shall show in the next section that if the Maxwell-Boltzmann velocity distribution is taken into account, the factors ($\frac{4}{3}$) in Eqs. (5-11) and (5-12a) and ($\frac{2}{3}$) in Eqs. (5-12b) and (5-13) must be replaced by $\sqrt{2}$ and ($\sqrt{2}/2$), respectively. Thus the collision frequencies in Eqs. (5-11), (5-12) and (5-13) are too low by about 5.5 percent.

EXERCISE A gas is composed of two different kinds of hard sphere molecules, A and B, of diameters σ_A and σ_B, moving with velocities U_A

MOLECULAR COLLISIONS

and U_B. The numbers of molecules per unit volume are n_A and n_B. Show that the number of collisions between a given A molecule and B molecules is, assuming $U_A > U_B$, and writing $\sigma_{AB} = \frac{1}{2}(\sigma_A + \sigma_B)$,

$$z_{AB} = n_B \pi \sigma_{AB}^2 \frac{3U_A^2 + U_B^2}{3U_A} \tag{5-14}$$

and the total collision frequency between all A and B molecules in unit volume is

$$Z_{AB} = n_A n_B \pi \sigma_{AB}^2 \frac{3U_A^2 + U_B^2}{3U_A} \tag{5-15}$$

Why does the factor $\frac{1}{2}$ in Eq. (5-12) not appear here? Do these equations reduce to Eqs. (5-1) and (5-2) if $U_A \gg U_B$?

Solution

$$dz_{AB} = dn_B \pi \sigma_{AB}^2 u_{\text{rel}}$$
$$dn_B = (n_B/4\pi) \sin\theta\, d\theta\, d\phi$$
$$u_{\text{rel}} = \sqrt{U_A^2 + U_B^2 - 2U_A U_B \cos\theta}$$
$$z_{AB} = (n_B/4\pi)(\pi\sigma_{AB}^2) \int_0^{2\pi} d\phi \int_0^\pi u_{\text{rel}} \sin\theta\, d\theta$$

Set $x = \cos\theta$ and $y = U_A^2 + U_B^2 - 2U_A U_B x$ in order to obtain z_{AB} (note that it is necessary to assume $U_A > U_B$, so that y remains positive throughout the integration).

$$z_{AB} = \frac{n_B}{2} \pi \sigma_{AB}^2 \frac{1}{2U_A U_B} \int_{(U_A - U_B)^2}^{(U_A + U_B)^2} y^{1/2}\, dy$$
$$= \frac{n_B}{2} \pi \sigma_{AB}^2 \frac{1}{2U_A U_B} \frac{2}{3}[(U_A + U_B)^3 - (U_A - U_B)^3]$$

EXERCISE Setting $U_A = \sqrt{\pi RT/M_A}$ and $U_B = \sqrt{\pi RT/M_B}$ in Eqs. (5-14) and (5-15) show that

$$z_{AB} = n_B \pi \sigma_{AB}^2 U_A \left(1 + \frac{1}{3}\rho\right) \tag{5-16}$$

where $\rho = M_A/M_B$. If A is hydrogen and B is xenon, what is the percentage error in the estimate made earlier in this chapter of the collision frequency of a hydrogen molecule with xenon atoms?

C. COLLISION FREQUENCY FOR MOLECULES MOVING WITH VELOCITIES GIVEN BY THE MAXWELL-BOLTZMANN DISTRIBUTION In a real gas the molecules move with a distribution of velocities, and an exact calculation of the collision frequency must take this into account. The detailed calculation here is somewhat frightening at first sight, but it is not difficult to carry through. A gas contains two kinds of molecules, A and B, there being n_A molecules of A and n_B molecules of B in unit volume. Let dn_A be the number of A molecules whose velocity components in the x-direction are between u_x and $u_x + du_x$, those in the y-direction are between u_y and $u_y + du_y$ and those in the z-direction are between u_z and $u_z + du_z$. According to Eq. (4-50)

$$dn_A = n_A \left(\frac{m_A}{2\pi kT}\right)^{3/2} \exp\left(\frac{-m_A(u_x^2 + u_y^2 + u_z^2)}{2kT}\right) du_x\, du_y\, du_z \tag{5-17}$$

where m_A is the mass of an A molecule. Similarly, if dn_B is the number of B molecules whose velocity components are in the range u_x' to $u_x' + du_x'$, u_y' to $u_y' + du_y'$, u_z' to $u_z' + du_z'$ then

$$dn_B = n_B \left(\frac{m_B}{2\pi kT}\right)^{3/2} \exp\left(\frac{-m_B(u_x'^2 + u_y'^2 + u_z'^2)}{2kT}\right) du_x'\, du_y'\, du_z' \tag{5-18}$$

From Eq. (5-3), the number of collisions between a particular A molecule moving in its range of velocities and the B molecules moving in their range is

$$dz_{AB} = dn_B \pi \sigma_{AB}^2 u_{\text{rel}} \tag{5-19}$$

and the number of collisions between all A molecules moving in their range and all B molecules moving in their range must be

$$dZ_{AB} = dn_A\, dn_B \pi \sigma_{AB}^2 u_{\text{rel}} \tag{5-20}$$

In Eqs. (5-19) and (5-20), u_{rel} is the relative velocity of the A and B molecules,

$$u_{\text{rel}} = [(u_x - u_x')^2 + (u_y - u_y')^2 + (u_z - u_z')^2]^{1/2} \tag{5-21}$$

and σ_{AB} is the usual collision diameter. In order to find the total number of collisions in unit time and unit volume between all A and B molecules, Eqs. (5-17), (5-18), and (5-21) must be substituted into Eq. (5-20) and the result must be integrated over all values of u_x, u_y, u_z, u'_x, u'_y, and u'_z between minus infinity and infinity.

$$Z_{AB} = \frac{1}{8} n_A n_B \pi \sigma_{AB}^2 \frac{(m_A m_B)^{3/2}}{(\pi kT)^3} \int\int\int\int\int\int$$
$$\exp\left(-\frac{[m_A(u_x^2 + u_y^2 + u_z^2) + m_B(u'^2_x + u'^2_y + u'^2_z)]}{2kT}\right)$$
$$u_{\text{rel}}\, du_x\, du_y\, du_z\, du'_x\, du'_y\, du'_z \quad (5\text{-}22)$$

This formidable-looking integral can be evaluated in a rather lengthy but straightforward fashion as shown in Appendix 5-1. The result that is obtained is

$$Z_{AB} = n_A n_B \pi \sigma_{AB}^2 (8kT/\pi\mu)^{1/2} \tag{5-23}$$

where μ is the reduced mass of the $A - B$ pair,

$$\mu = \frac{m_A m_B}{m_A + m_B} \tag{5-24}$$

If A and B are identical, then we may write $m_A = m_B = m$, $\mu = m/2$, $\bar{u}_A = \bar{u}_B = \bar{u}$, $n_A = n_B = n$, $\sigma_{AB} = \sigma$ and we find for the collision frequency in unit volume of a pure gas

$$Z = \frac{1}{2} n^2 \pi \sigma^2 [8kT/\pi(m/2)]^{1/2} = \frac{\sqrt{2}}{2} n^2 \pi \sigma^2 (8kT/\pi m)^{1/2}$$
$$= \frac{\sqrt{2}}{2} n^2 \pi \sigma^2 \bar{u} \tag{5-25}$$

a new factor of $\frac{1}{2}$ being introduced because when A and B are the same, Eq. (5-23) counts each collision twice, as explained previously. Equation (5-25) is the final, correct expression for the collision frequency in a gas containing a single kind of molecule. It is of the same form as Eq. (5-13) except that the factor $\frac{2}{3}$ in the latter has been replaced by the factor $\sqrt{2}/2$.

The number of collisions per second taking place between a given A molecule with B molecules is found by dividing Eq. (5-23) by n_A,

$$z_A = n_B \pi \sigma_{AB}^2 (8kT/\pi\mu)^{1/2} \tag{5-26}$$

and if A and B are identical this becomes

$$z = \sqrt{2}\, n\pi\sigma^2 \bar{u} \tag{5-27}$$

which is the same as Eq. (5-11) if in that equation we write $\bar{u}$ for u and if the factor $\frac{4}{3}$ is replaced by $\sqrt{2}$.

The total number of collisions between a given A molecule and all other kinds of molecules in a mixture containing more than two substances is

$$z_A = \sum_i n_i \pi \sigma_{iA}^2 (8kT/\pi\mu_{iA})^{1/2} \tag{5-28}$$

where n_i is the number of molecules of type i present in unit volume, σ_{iA} is the collision diameter for A molecules with i molecules and μ_{iA} is the reduced mass of molecules A and i ($\mu_{iA} = m_A m_i / (m_A + m_i)$). If molecule A is very light compared with all other molecules then $\mu_{iA} \cong m_A$ and we find

$$z_A \cong \pi \bar{u}_A \sum_i n_i \sigma_{iA}^2 \tag{5-29}$$

If the gas consists only of light molecules A and heavy ones B Eq. (5-29) becomes

$$z_A \cong \pi n_B \sigma_{AB}^2 \bar{u}_A \tag{5-30}$$

which is identical with Eq. (5-1) if the velocity U in that equation is replaced by $\bar{u}_A$. Thus the assumption that all light molecules move with the same velocity, U, gives the correct collision frequency if the velocity U is replaced by the mean thermal velocity of the light molecules. Equations (5-29) and (5-30) are suitable for calculating the collision frequencies of electrons moving randomly through a gas.

d. COLLISION FREQUENCY OF GAS MOLECULES WITH A SURFACE
Consider a surface of area dS exposed to a gas containing n molecules per unit volume moving at an average velocity $\bar{u}$. Consider the molecules whose directions of motion lie in a solid angle $d\Omega = \sin\theta\, d\theta\, d\phi$ defined by the range of polar angles θ, $\theta + d\theta$, ϕ, $\phi + d\phi$, where θ is the angle between the normal to dS and the direction of molecular motion. The number of such molecules per unit volume is $(n/4\pi)\, d\Omega$ (see Eq. (5-9)). The component of molecular velocity normal to dS is $\bar{u} \cos\theta$, and the number of molecules in $d\Omega$ colliding with dS in a time interval dt is equal to the number of molecules contained in a cylinder of base dS and

altitude $\bar{u}\cos\theta\,dt$, that is

$$dn = (n/4\pi)\,d\Omega \times dS \times \bar{u}\cos\theta\,dt$$
$$= (\bar{u}n/4\pi)\,dS\,dt\,\sin\theta\cos\theta\,d\theta\,d\phi \quad (5\text{-}31)$$

Integrating over all directions ϕ from 0 to 2π, and allowing θ to go from 0 to $\pi/2$ (since molecules can approach the surface dS from one side only), we find for the collision number per unit time and per unit area of surface

$$z_{\text{surf}} = \frac{1}{dS\,dt} \int dn$$
$$= (\bar{u}n/4\pi) \int_0^{2\pi} d\phi \int_0^{\pi/2} \cos\theta \sin\theta\,d\theta$$
$$= \frac{1}{4}n\bar{u} \quad (5\text{-}32)$$

EXERCISE A small "oven" with a 1 mm² hole in its side and containing several grams of mercury at 20°C is placed in a high vacuum under continuous pumping. How long will it take for 1 mg of mercury to effuse from the oven? (The vapor pressure of mercury is 0.00120 torr at 20°C. Assume that mercury vapor is monatomic.)

EXERCISE Stefan's law states that in an evacuated space enclosed by a container whose walls are at a temperature T, the energy density of the thermal radiation is

$$\epsilon = aT^4 \quad (5\text{-}33)$$

where a is the Stefan-Boltzmann constant, $a = 7.57 \times 10^{-15}$ erg-cm^{-3} (°K)$^{-4}$. The thermal radiation moves back and forth in the box in random directions with the speed of light, $c = 2.99 \times 10^{10}$ cm/sec. If a hole of area dS is present in the container, show that the rate of emission of thermal radiation from this hole ("black body radiation") is given by

$$q = \frac{1}{4}caT^4\,dS \quad (5\text{-}34)$$

This is the maximum rate of radiation of thermal energy from a surface at temperature T. Real surfaces emit at a rate

$$q = \frac{1}{4}ecaT^4\,dS = 5.67 \times 10^{-5} eT^4\,dS \text{ ergs/sec}$$
$$= 1.36 \times 10^{-12} eT^4\,dS \text{ cal/sec} \quad (5\text{-}35)$$

where e is a number lying between 0 and 1, called the emissivity of the surface. A surface with $e = 1$ is called a "black" surface. If $T = 100°K$, then $q = 1.36 \times 10^{-4}$ cal/sec, and if $T = 1000°K$, $q = 1.36$ cal/sec for one square centimeter of black surface. Thus thermal radiation is rather slight at low temperatures but becomes considerable for hot objects.

5-2 The mean free path

The mean free path is defined as the mean distance travelled by a molecule between two successive collisions. For a pure gas, from Eq. (5-27), the number of collisions suffered by a given molecule in one second is $\sqrt{2}\, n\pi\sigma^2 \bar{u}$, and since the average distance travelled by the molecule in one second is $\bar{u}$, the average distance between collisions—that is, the mean free path, λ—must be

$$\lambda = \frac{\bar{u}}{z} = \frac{1}{\sqrt{2}\, n\pi\sigma^2} \tag{5-36}$$

In a mixture of gases, the mean free path of a molecule of type A is

$$\lambda_A = \frac{\bar{u}_A}{z_A} \tag{5-37}$$

where z_A is given by Eq. (5-28) and $\bar{u}_A = (8kT/\pi m_A)^{1/2}$. Substituting for the masses of the different types of gases we find

$$\lambda_A = \frac{1}{\sum_i n_i \pi \sigma_{iA}^2 [1 + (m_A/m_i)]^{1/2}} \tag{5-38}$$

If the molecules A are very light in comparison with the other molecules, the mean free path of the light molecules is

$$\lambda_A = \frac{1}{\sum n_i \pi \sigma_{iA}^2} \tag{5-39}$$

an expression suitable for calculating the mean free path of an electron in a gas.

According to the above relations, the mean free path of a molecule in a gas composed of rigid molecules should be independent of the temperature at constant density, but if the temperature is fixed the mean free path will depend on the density and hence on the pressure. For air ($\sigma = 3.8 \times 10^{-8}$ cm) at 1 atm and 300°K ($n = 2.45 \times 10^{19}$ molecules/cm^3), we find from Eq. (5-36) that the mean free path is $1/\sqrt{2} \times (2.45 \times 10^{19}$ molecules/cm$^3) \times (3.8 \times 10^{-8}$ cm$)^2 = 2.0 \times 10^{-5}$ cm, which is about 500 molecular diameters. This justifies the assumption underlying the calculation of the collision frequency, that the collision frequency can be found from the volume of the straight cylinder swept out in unit time by a disc of area $\pi\sigma^2$ and length u_{rel}. Obviously the kinks in the cylinder caused by collisions require a correction of the order of 0.5% at one atmosphere; at lower pressures the correction is proportionately smaller. It is also obvious that at pressures of the order of 100 atm or more the simple theory will be seriously in error.

Since the number of molecules in unit volume of an ideal gas varies in proportion to the pressure, we may write for the mean free path at pressure p

$$\lambda_p = \frac{\lambda_1}{p} \tag{5-40}$$

where λ_1 is the mean free path when $p = 1$. Since $\lambda_1 \cong 10^{-5}$ cm at one atmosphere for simple gases at room temperature, mean free paths will be of the order of centimeters or more at pressures of 10^{-5} atm (10^{-2} mm Hg, 10^{-2} torr or 10 μ) and below. At such pressures we may expect that in vessels and tubing of the dimensions commonly used in laboratory equipment, molecules will collide with the walls more frequently than they collide with each other. Properties of gases that depend on molecular collisions (viscosity, heat conduction, diffusion) may therefore be expected to undergo marked changes at these low pressures.

EXERCISE The density of gas in interstellar space is believed to be of the order of 1 molecule per cubic centimeter, and the interstellar gas is largely hydrogen. If $\sigma = 1 \times 10^{-8}$ cm, what is the mean free path in interstellar space? What is the collision frequency of a given hydrogen molecule (in collisions per century) if the temperature of interstellar space is 10°K?

The definition of the mean free path given above is only one of various possible ways of defining an average path for a molecule between collisions. Since the collision frequency depends on the relative velocity of the molecules involved in a collision, it is not

TABLE 5-1 DEPENDENCE OF MEAN FREE PATH ON MOLECULAR VELOCITY

$u/\bar{u}$	0	0.25	0.5	1.0	2.0	4.0	∞
$\lambda_u/\lambda_0{}^a$	0	0.3445	0.6411	1.0257	1.2878	1.3803	1.414

^{a} $\lambda_0 = 1/\sqrt{2}\,\pi\sigma^2 n$ = mean free path averaged over all velocities.

surprising that the mean free path of a given molecule depends on its velocity. (The mean free path of a stationary molecule is zero, whereas comparison of Eqs. (5-36) and (5-39) shows that a molecule moving at a very large velocity relative to other molecules in a gas has a mean free path which is $\sqrt{2}$ times greater than the mean free path averaged over all molecules.) It is shown in the exercise below that in a pure gas a molecule moving with velocity u has a mean free path

$$\lambda_u = \frac{mu^2}{2\sqrt{\pi}\,n\sigma^2 \psi(u/u_0)} \tag{5-41}$$

where $u_0 = (2kT/m)^{1/2}$, the most probable velocity, and

$$\psi(x) = x\exp(-x^2) + (2x^2 + 1)\int_0^x \exp(-y^2)\,dy \tag{5-42}$$

Table 5-1 gives values of λ_u/λ_0 for various values of $u/\bar{u}$ where λ_u is calculated from Eqs. (5-41) and (5-42), λ_0 is calculated from Eq. (5-36) and $\bar{u}$ is the mean thermal velocity. A strong dependence of the mean free path on molecular velocity is evident, especially at low velocities.

EXERCISE Prove Eqs. (5-41) and (5-42). [*Hint:* Let a given molecule move with a velocity u. Let the direction of this motion define the polar axis of a polar coordinate system (θ, ϕ). The number of other molecules per unit volume moving with velocities in the range u', $u' + du'$ in directions defined by the range of polar angles θ, $\theta + d\theta$, ϕ, $\phi + d\phi$ is

$$dn' = n\left(\frac{m}{2\pi kT}\right)^{3/2}\exp\left(-\frac{mu'^2}{2kT}\right)u'^2 \sin\theta\,d\theta\,d\phi\,du'$$

The collision frequency dz' between these molecules and the given molecule is found by multiplying dn' by $\pi\sigma^2 u_{\text{rel}}$, where u_{rel} is given by

$$u_{\text{rel}}^2 = u^2 + u'^2 - 2uu'\cos\theta$$

MOLECULAR COLLISIONS

For constant u and u' we have

$$uu' \sin\theta \, d\theta = u_{\text{rel}} \, du_{\text{rel}}$$

so the collision frequency dz' is

$$dz' = \pi\sigma^2 n \left(\frac{m}{2\pi kT}\right)^{3/2} \exp\left(-\frac{mu'^2}{2kT}\right) \frac{u'}{u} \, du' \, u_{\text{rel}}^2 \, du_{\text{rel}} \, d\phi$$

This can be integrated over ϕ to give a factor of 2π. The integration over u_{rel} is between the limits $u - u'$ to $u + u'$, if $u > u'$, or between the limits $u' - u$ and $u + u'$ if $u < u'$. Finally, the integration over u' may be performed, giving for the collision frequency between a molecule with velocity u and all molecules moving with any other velocity in the Maxwell-Boltzmann distribution

$$z_u = \frac{4}{3\sqrt{\pi}} n\pi\sigma^2 \left(\frac{m}{2kT}\right)^{3/2} \left[\int_0^u \frac{u'^2(u'^2 + 3u^2)}{u} \exp\left(-\frac{mu'^2}{2kT}\right) du' \right.$$
$$\left. + \int_u^\infty u'(u^2 + 3u'^2) \exp\left(-\frac{mu'^2}{2kT}\right) du'\right]$$

Evaluating these integrals and writing $\lambda_u = u/z_u$ gives Eqs. (5-41) and (5-42).]

Another problem closely related to the mean free path arises from the obvious fact that the distance travelled by a given molecule between collisions must vary randomly in successive collisions. Let $P_u(l)$ be the probability that a molecule travelling with velocity u moves through a distance l without experiencing a collision with another molecule. The probability that this molecule will undergo a collision while moving through a short distance dl is dl/λ_u. This probability, when multiplied by $P_u(l)$, gives us $-dP_u(l)$, the decrease in $P_u(l)$ resulting from motion through the distance dl. That is,

$$dP_u(l) = -\frac{dl}{\lambda_u} P_u(l) \tag{5-43}$$

which on integration, using the initial condition $P_u(0) = 1$, gives

$$P_u(l) = e^{-l/\lambda_u} \tag{5-44}$$

This is the distribution function for free path lengths. From it we may confirm that the mean free path for molecules of velocity

u is λ_u

$$\overline{\lambda_u} = \int_0^\infty lP_u(l)\,dl \Big/ \int_0^\infty P_u(l)\,dl$$

$$= \int_0^\infty le^{-l/\lambda_u}\,dl \Big/ \int_0^\infty e^{-l/\lambda_u}\,dl$$

$$= \lambda_u{}^2 \int_0^\infty xe^{-x}\,dx \Big/ \lambda_u \int_0^\infty e^{-x}\,dx = \lambda_u \qquad (5\text{-}45)$$

This distribution is fairly broad; the fraction of molecules with free paths less than $0.5\lambda_u$ is $1 - e^{-0.5} = 0.394$, and the fraction with free paths greater than $2\lambda_u$ is $e^{-2} = 0.136$.

Equation (5-44) has been tested experimentally by investigating the attenuation of beams of electrons or molecules passing through a gas. Such experiments make it possible to measure the mean free path and thus the collision diameter.

5-3 The transport properties of gases

If the properties of a gas vary from one point to another, then in general a flow of some associated quantities will take place through the gas. For instance, if the temperature of a gas is not uniform, *heat conduction* will take place from regions of high temperature to regions of low temperature. If the mean mass velocity of the gas is not everywhere the same, forces will be exerted by each portion of the gas on adjacent portions tending to eliminate these velocity differences—a process giving rise to the phenomenon of *viscosity*. If the composition is not everywhere the same, then the process of *diffusion* will take place. Such phenomena are known as *transport processes*. In this section we shall describe the phenomenological characteristics of several kinds of transport processes, and in the remainder of the chapter the molecular basis for these processes will be considered.

a. HEAT CONDUCTION Let there be two parallel plates of area A, separated by a distance d, and let the plates differ in temperature by ΔT. Then it is observed that if a substance is present between the plates, and if convection is avoided in the substance, heat will flow through the substance at the rate

$$q = \frac{\kappa A\,\Delta T}{d} \qquad (5\text{-}46)$$

TABLE 5-2 TYPICAL VALUES OF HEAT CONDUCTIVITIES

Substance	Temp	κ(cal/deg cm)
Copper	18°C	0.92
Iron	18°C	0.16
Glass	20°C	0.002
Wood		0.0002
Water (liquid)	17°C	0.0013
Benzene (liquid)	5°C	0.0003
Air	0°C	0.000057
Hydrogen (gas)	0°C	0.00042
Helium (gas)	0°C	0.00034
Radiation across 1 cm gap[a]	100°K	0.0000055
	300°K	0.00015
	1,000°K	0.0055
	10,000°K	5.5

[a] See exercise at the bottom of this page.

where κ is a property of the substance known as the *thermal conductivity*. If q is measured in calories per second and the dimensions A and d are measured in centimeters, then κ has the units cal/cm deg sec. The constant κ gives the rate of heat flow through a cube of substance measuring 1 cm on each edge, and differing in temperature by 1° on a pair of opposite faces. Table 5-2 gives the values of the constant κ for a number of familiar substances. It is evident that gases are relatively poor conductors of heat whereas metals are very good heat conductors.

EXERCISE If two plates of area A and of temperatures T and $T + \Delta T$ are separated by an evacuated space, each will emit thermal radiation at a rate given by Eq. (5-35). Show that the net conduction of heat from the hot plate to the cooler one is, assuming $\Delta T \ll T$,

$$q = caeAT^3 \Delta T$$
$$= 5.4 \times 10^{-12} eAT^3 \Delta T \text{ cal/sec} \tag{5-47}$$

where e is the emissivity of the plates, c is the velocity of light, a is the Stefan-Boltzmann constant, A is expressed in cm^2, and T is in degrees Kelvin. Note that the heat flux here is independent of the separation of the plates; this assumes that the dimensions of the plates are large compared with their separation. The magnitudes of the rates of heat transfer through a vacuum in Table 5-2 are obtained from Eq. (5-47) with $A = 1$ cm^2, $e = 1$, and $\Delta T = 1°$C. A vacuum will "conduct" heat about as well as a gas does if the walls are at 300°K unless the walls are coated with a material of low emissivity. This is the reason for silvering the walls of Dewar flasks. At 1000°K a vacuum is a fairly good transporter of heat, and at 10,000°K it transports heat over distances of the order of 1 cm about as well as a metal does at room temperature.

The thermal conductivity of a gas is usually measured by either of two methods: the *plate method* and the *hot wire method*. In the plate method the gas is placed between two concentric cylindrical surfaces—an inner cylinder which can be of solid metal, and an outer can which is held at a fixed temperature by immersion in a constant temperature bath. The space between the inner and outer cylinders contains the gas and must be sufficiently small so that convection does not take place. (Convection can also be avoided by working at reduced pressures, but the pressure must not be so low that the mean free path of the gas becomes comparable with the space between the cylinders.) The interior cylinder is warmed to a temperature T_1 and its temperature is then measured as a function of the time as it cools toward the temperature T_0 of the surroundings. If the heat capacity of the inner cylinder is C, and if the average area of the cylinders is A, and if the spacing between the cylinders is d, then a change dT in the temperature of the inner cylinder involves a heat loss $-dq = C\, dT$ by this cylinder. This heat loss is accomplished in part by the heat conduction through the gas; in a time interval dt this contribution is

$$dq_{\text{gas}} = \frac{\kappa A (T - T_0)}{d}\, dt \tag{5-48}$$

where T_0 is the temperature of the outer cylinder. Some heat loss also takes place by radiation and by conduction through the mechanical supports of the inner cylinder. Generally these heat losses (which may be termed "apparatus losses") occur at a rate proportional to $T - T_0$ so we may write

$$dq_{\text{app}} = \alpha_{\text{app}}(T - T_0)\, dt \tag{5-49}$$

where α_{app} is a constant, characteristic of the apparatus. Thus we may write

$$dq = -C\,dT$$
$$= dq_{gas} + dq_{app}$$
$$= (\alpha_{gas} + \alpha_{app})(T - T_0)\,dt \quad (5\text{-}50)$$

where we have written

$$\alpha_{gas} = \frac{\kappa A}{Cd} \quad (5\text{-}51)$$

We thus obtain the differential equation

$$\frac{dT}{T - T_0} = -(\alpha_{gas} + \alpha_{app})\,dt \quad (5\text{-}52)$$

which has the solution

$$\ln(T - T_0) = -(\alpha_{gas} + \alpha_{app})t \quad (5\text{-}53)$$

A plot of $\ln(T - T_0)$ against the time, t, should give a straight line from whose slope $(\alpha_{gas} + \alpha_{app})$ may be determined. The apparatus constant α_{app} may be determined by pumping away all of the gas and repeating the experiment. Thus the value of α_{gas} may be found and if the apparatus characteristics C, A, and d are known, the heat conductivity κ of the gas may be measured.

In the hot wire method of measuring the heat conductivity of a gas, a platinum wire is stretched along the axis of a cylindrical tube containing the gas. The wire is heated with a known current, i, at a measured voltage, v. The rate of supply of energy to the wire is

$$q = iv \quad (5\text{-}54)$$

On reaching a steady state this input is balanced by the heat conduction, which can be shown to be

$$q = \frac{2\pi\kappa l(T_2 - T_1)}{\ln(r_1/r_2)} + q_r \quad (5\text{-}55)$$

where l is the length of the wire, T_2 is the temperature of the platinum wire (determined by comparing its electrical resistance, v/i, at T_2 to its resistance at a known temperature), T_1 is the temperature of the outside tube, r_1 is the radius of the tube, r_2 is the radius of the platinum wire, and q_r is the transport by radiation (determined by experiments in which the cylinder is evacuated). The heat input being known from Eq. (5-54), κ can be found from Eq. (5-55).

b. VISCOSITY Consider two parallel plates of area A separated by a distance d, which is small compared with the dimensions of the plates. One of the plates is stationary and the other moves with a velocity U parallel to the stationary plate (Fig. 5-4a). If a liquid or gas fills the space between the plates, then in order to maintain the velocity of the moving plate at a constant value, a force f must be applied to it; this force acts in the same direction as the direction of motion U. It is observed that the force f is proportional to the velocity U and to the area of the plates, A, and inversely proportional to the separation of the plates d

$$f = \frac{\eta A U}{d} \tag{5-56}$$

where η is a constant of proportionality whose magnitude is a property of the substance between the plates; η is known as the *viscosity* of the substance. Equation (5-56) is sometimes called Newton's law of flow, because Newton first noted the proportionality between V and f. The viscosity η is commonly expressed in units of dynes/cm^2 sec—a unit also known as the *poise*. The number η therefore gives the shearing stress in dynes per cm^2 required to maintain a difference in velocity of 1 cm/sec across a 1 cm layer of a substance. Typical values of η for various fluids are shown in Table 5-3. Gases are very much less viscous than the common liquids such as water and benzene. It is interesting to note that among the gases themselves there is considerably less variation in the viscosity than there is for the heat conduction.

It should be mentioned that the proportionality between the shear stress f and the shear velocity gradient U/d expressed in Eq. (5-56) is not observed for solids. There are also complex liquid mixtures which fail to obey Eq. (5-56), and the equation fails for certain materials at high shear velocities. Materials which fail to obey Eq. (5-56) are said to be *non-Newtonian*.

Three methods are commonly used for measuring the viscosities of gases: the capillary flow method, the concentric cylinder method and the damped oscillation method. The capillary flow method

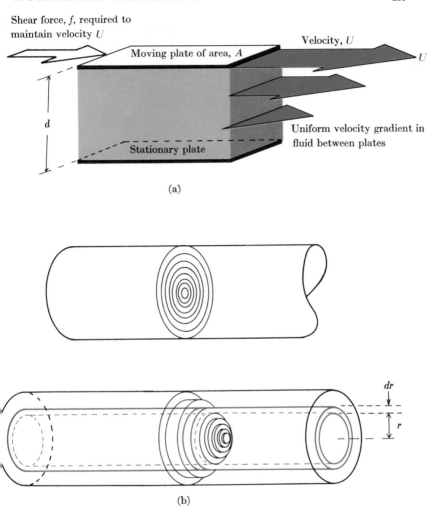

FIG. 5-4 *Viscous flow of a fluid.* (a) *Flow accompanying the shearing motion of two parallel plates.* (b) *Viscous flow of a fluid through a tube. Layers of liquid shear past each other as if they were tubes of a collapsible telescope. The upper sketch shows a cross section at some instant, imagined to have been divided into concentric tubes. The lower sketch shows the same cross section at a later time, the tubes of fluid having pushed forward at different velocities. The forces on the tube that has been outlined in color in the lower sketch are considered in the derivation of Poiseuille's law in the exercise that follows Eq.* (5-57).

TABLE 5-3 VISCOSITIES OF VARIOUS FLUIDS

Substance	Viscosity (poises)
Liquids (20°C)	
n-Hexane	0.0033
Benzene	0.0065
Water	0.0100
Olive oil	0.84
Glycerol	15.
Gases (0°C)	
Hydrogen	8.4×10^{-5}
Helium	18.4×10^{-5}
Air	17.1×10^{-5}
Water	8.6×10^{-5}
Benzene	6.8×10^{-5}
n-Hexane	5.9×10^{-5}

depends on Poiseuille's law for the flow of a Newtonian fluid through a tube,

$$q = \frac{\pi R^4 \, \Delta p}{8\eta l} \tag{5-57}$$

where q is the volume flow rate of the fluid, R is the radius of the tube, Δp is the pressure difference between the two ends of the tube and l is the length of the tube. In this method the experimenter measures the time required for a known volume of gas to flow through a tube under a measured pressure difference Δp. The proportionality constant, $\pi R^4/8l$, between q and $\Delta p/\eta$ can be determined either by measuring the dimensions of the tube (which is usually not very easy to do since fine capillaries are employed) or by measuring q and Δp for a fluid of known viscosity.

EXERCISE Derive Poiseuille's law. [*Hint:* Assume that the flow through the tube takes place by concentric cylindrical layers of fluid slipping past one another, as in a collapsible telescope. Consider one of these tubes, whose inner surface has the radius r, and whose outer surface has the radius $r + dr$ (Fig. 5-4b). Since the pressure difference at the two ends of the tube is Δp, and the area of the (plane) ends of the cylindrical tube is πr^2, so that a force difference $\Delta p \pi r^2$ acts on the two ends, and since the area of the cylindrical surface of the tube is $2\pi r l$, there must be a shear stress $\Delta p \pi r^2 / 2\pi r l = \Delta p r / 2l$ acting across the inner and outer surfaces of the tube. This stress causes the inner and outer surfaces of the tube to move with different velocities, $u(r) - u(r + dr) = -du = (\Delta p r / 2l\eta) \, dr$, $u(r) = \text{const} - (\Delta p / 4l\eta) r^2$. It is reasonable to assume as a boundary condition that the layer of fluid next to the wall of the capillary (at $r = R$) is stationary—that is, $u(R) = 0$, so that $u(r) = (\Delta p / 4\eta)(R^2 - r^2)$. The total flow rate through the capillary is then

$$q = \int_0^R u(r) 2\pi r \, dr \tag{5-58}$$

For gases flowing through very fine capillaries it appears to be necessary to make a correction for "slip" of the gas at the walls of the tube, arising from the failure of the boundary condition, $u(R) = 0$, in the derivation of the Poiseuille equation.]

In the rotating cylinder method of measuring the viscosity, the fluid is introduced between two concentric cylinders, the separation of the cylinders being small compared with the radii of the cylinders. One of the cylinders is rotated at a constant rate and the other is mounted on a torsion wire whose twist is a measure of the viscous force transmitted through the fluid.

In the damped oscillation method a set of parallel, horizontal vanes is suspended from a torsion wire. Interleaving with these vanes is a set of fixed vanes. The fluid whose viscosity is to be measured fills the space between the fixed and movable vanes. The movable vanes are set in oscillation and the rate of damping of these oscillations is observed. The damping rate is proportional to the viscosity of the fluid. The apparatus may be calibrated with a fluid of known viscosity.

C. DIFFUSION Let two mutually soluble substances, 1 and 2, be mixed in such a way that the composition is different at different points in the mixture. Then experience tells us that over a period of time substances 1 and 2 will move from regions in which their concentrations are high into regions where their concentrations are low. This process, known as diffusion, continues until the composition of the mixture is everywhere the same. It is observed that the flow of a substance by diffusion across an arbitrary plane in a mixture is proportional to, and opposite in direction to, the concentration gradient normal to the plane

$$j_1 = -D_{12}A\frac{dc_1}{dz} = -D_{12}A\rho\frac{dx_1}{dz}$$
$$j_2 = -D_{21}A\frac{dc_2}{dz} = -D_{21}A\rho\frac{dx_2}{dz} \quad (5\text{-}59)$$

where j_1 is the flux of component 1—that is, the quantity flowing per unit time (say in moles per second) through the area A of the plane, c_1 is the concentration of component 1, ρ is the density of the mixture (say in moles/cm^3), x_1 is the mole fraction of 1 in the mixture, z is the distance measured normal to the plane and D_{12} is a property of the mixture known as the *diffusion coefficient* or *diffusivity* of substance 1 through substance 2.

The quantities j_2, c_2, x_2, and D_{21} are similarly defined for substance 2. This proportionality of diffusion flux to concentration gradient is sometimes called Fick's first law of diffusion.

It is useful to investigate the time variation of the concentration at a given point in space caused by diffusion. Consider a thin slab of the mixture of thickness dz and area A. Let one face of this slab have the coordinate z, the z-axis being perpendicular to the slab, so that the other face is at $z + dz$. The rate of flow of substance 1 in the z-direction through the face of the slab at z is, by Fick's first law

$$(j_1)_z = -D_{12}A\frac{dc_1}{dz} \quad (5\text{-}60)$$

and the flow through the face at $z + dz$ is

$$(j_1)_{z+dz} = -\left[(j_1)_z + A\frac{d}{dz}\left(D_{12}\frac{dc_1}{dz}\right)dz_1\right] \quad (5\text{-}61)$$

In a time interval dt, the net accumulation of substance 1 in the slab is

$$dm_1 = [(j_1)_z - (j_1)_{z+dz}]\, dt$$
$$= A\, dz\, \frac{d}{dz}\left(D_{12}\frac{dc_1}{dz}\right) dt \quad (5\text{-}62)$$

Since the volume of the slab is $V = A\, dz$ and since $c_1 = m_1/V$, we have

$$\frac{dc_1}{dt} = \frac{d}{dz}\left(D_{12}\frac{dc_1}{dz}\right) \quad (5\text{-}63)$$

It is important to note that the derivative on the left side of this equation is taken at a fixed location in the mixture (i.e., at constant z) whereas the derivatives on the right side are taken at a fixed instant in time. This may be indicated by using the notation of partial differentiation (see Chapter 1 of Volume II)

$$\left(\frac{\partial c_1}{\partial t}\right)_z = \left[\frac{\partial}{\partial z}\left(D_{12}\frac{\partial c_1}{\partial z}\right)\right]_t \quad (5\text{-}64)$$

a relationship known as Fick's second law of diffusion. The diffusivity D_{12} may in general vary with the composition and may therefore be a function of the location z in the mixture. If this variation is neglected, however, we can write

$$\left(\frac{\partial c_1}{\partial t}\right)_z = D_{12}\left(\frac{\partial^2 c_1}{\partial z^2}\right)_t \quad (5\text{-}65)$$

Consider the experimental situation in which a thin layer of substance 1 is introduced in substance 2 at time $t = 0$. Then direct substitution into Eq. (5-65) shows that the subsequent diffusion of substance 1 follows the relation

$$c(z, t) = \frac{c_0}{\sqrt{4\pi Dt}} \exp\left(\frac{-z^2}{4Dt}\right) \quad (5\text{-}66)$$

where c_0 is the amount of substance 1 introduced in unit area of interface, z is the distance from the original position of the layer and D is the diffusion coefficient for substance 1 into substance 2. Profiles of the concentration versus z are shown in Fig. 5-5, the time being measured in units of $\tfrac{1}{4}D$. The mean square distance moved at time t from the initial layer is

$$\overline{z^2} = \int_{-\infty}^{\infty} z^2 c\, dz \Big/ \int_{-\infty}^{\infty} c\, dz$$

$$= \int_{-\infty}^{\infty} z^2 \exp\left(\frac{-z^2}{4Dt}\right) dz \Big/ \int_{-\infty}^{\infty} \exp\left(\frac{-z^2}{4Dt}\right) dz$$

$$= \frac{\sqrt{\pi}}{2}(4Dt)^{3/2} \Big/ \sqrt{4\pi Dt} = 2Dt \quad (5\text{-}67)$$

The quantity $\sqrt{\overline{z^2}}$ is a measure of the distance traveled by diffusion; evidently this distance varies with the square root of the time. Thus motion by means of diffusion is quite unlike ordinary uniform motion in that, if it requires a time t_0 for an average molecule to travel a distance z_0, then it requires a time $4t_0$ for the average molecule to travel a distance $2z_0$.

The diffusion coefficient itself gives the inverse of the time required to travel a root mean square distance $\sqrt{2}$. Typical values of diffusion coefficients are given in Table 5-4. In gases the typical molecule moves a distance of the order of 1 cm in a few seconds, whereas small molecules in aqueous solutions require roughly a day to diffuse a root mean square distance of 1 cm. A protein molecule requires several weeks to diffuse this far in water, on the average. Diffusion in solids at room temperature is an even slower process, but at elevated temperatures diffusion in solids can be quite rapid. Particularly interesting for theoretical purposes

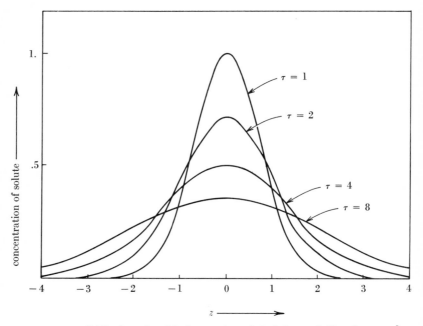

FIG. 5-5 *Diffusion of a thin layer of a solute into a solution in accordance with Eq. (5-65). Time is measured in units of $\tau = (\text{unit length})^2/4D$.*

TABLE 5-4 DIFFUSION COEFFICIENTS AND THERMAL DIFFUSION FACTORS OF VARIOUS MIXTURES

Mixture	Temp (°C)	D(cm²/sec)	α
Gases (1 atm)			
p-H_2-H_2	0°	1.285	—
D_2-H_2	0	1.13	0.17
Ne^{20}-Ne^{22}	0	0.452	0.019
A^{36}-A^{40}	0	0.158	0.10
Xe-Xe	27	0.0576	—
Ne-A	0	—	0.15
He-A	0	0.64	0.38
O_2-N_2	0	0.18	—
Liquids			
D_2O-H_2O	0	1×10^{-5}	—
	25	2.1×10^{-5}	—
H_2O^{18}-H_2O^{16}	0	1.5×10^{-5}	—
	25	3.0×10^{-5}	—
NaCl-H_2O	25	1.9×10^{-5}	—
Sucrose-H_2O	25	0.52×10^{-5}	—
Serum albumin-H_2O	25	0.059×10^{-5}	—
Solids			
Pb in Pb	25	10^{-20a}	—
	100	10^{-16}	—
	300	2×10^{-10}	—
Au in Pb	25	10^{-10a}	—
	100	5×10^{-9}	—
	300	10^{-6}	—
Au in Au	25	10^{-24a}	—
	100	3×10^{-21a}	—
	300	2×10^{-15}	—

a Extrapolated from measurements at higher temperatures.

are the diffusion coefficients in mixtures of two isotopic species of the same substance, such as H_2O^{18} in H_2O^{16} or $C^{14}H_4$ in $C^{12}H_4$. These are called *self-diffusion coefficients*.

EXERCISE If the diffusing substances are ideal gases, if c_1 and c_2 are molar concentrations and if the pressure and temperature are uniform throughout the system, then $c_1 + c_2$ must be everywhere the same. Show that under these conditions $D_{12} = D_{21}$.

d. THERMAL DIFFUSION If a mixture of gases is placed in a tube and the two ends of the tube are maintained at different temperatures, then it is observed that a steady state is eventually reached in which the composition of the gas varies along the tube. Evidently, the ordinary diffusion of the gases down their concentration gradients is opposed by a "thermal diffusion" process which tends to make the molecules move along the temperature gradient. The existence of this phenomenon was predicted theoretically and independently in 1917 by Enskog and by Chapman. The phenomenon had been used to assist in the separation of isotopes.

Consider a mixture of two substances, 1 and 2, which is subjected to a temperature gradient dT/dz, which in turn produces a concentration gradient dx_1/dz, where x_1 is the mole fraction of component 1. Then if ρ is the density of the gas, in moles per milliliter, we may write, when the steady state has been achieved

$$D_{12}\rho \frac{dx_1}{dz} = D_T \rho \frac{d \ln T}{dz} \tag{5-68}$$

where the left-hand side of the equation gives the ordinary diffusion rate and the right-hand side gives the thermal diffusion rate. The quantity D_T is called the *coefficient of thermal diffusion* and because it is defined in terms of the derivative of the logarithm of the temperature, it has the same dimensions as D_{12} (i.e., cm^2/sec). It is found that D_T is almost invariably positive for the flow of the lighter component of a mixture through the heavier one (i.e., if dT/dz is positive, then dx/dz is positive for the light component of a mixture and negative for the heavy component; thus the light component usually moves toward the higher temperature). The numerical value of D_T for a mixture is seldom as large as 30% of D_{12} for the mixture and it is usually smaller than this.

Equation (5-18) may be written in the form

$$\frac{dx_1}{dz} = \frac{D_T}{D_{12}} \frac{d \ln T}{dz} \tag{5-69}$$

It is found experimentally that for a given pair of gases the ratio D_T/D_{12} varies with the composition in such a way that the

MOLECULAR COLLISIONS

quantity

$$\alpha = \frac{D_T/D_{12}}{x_1 x_2} \tag{5-70}$$

is independent of composition where x_1 and x_2 are the mole fractions of the components of the mixture. The quantity α is called the *thermal diffusion factor*, and it is usually given a sign corresponding to the light component of the mixture moving to the higher temperature.

Some typical values of the thermal diffusion factor, α, are given in Table 5-4.

EXERCISE A mixture of H_2 and CO_2 is exposed to a temperature gradient of 230°C at the hot end of a tube and 10°C at the cold end. When a steady state has been reached it is found that at the hot end of the tube the gas contains 44.9% H_2, whereas at the cold end it contains 41.3% H_2. Estimate the values of D_T/D_{12} and α for this mixture.

5-4 Approximate molecular theory of transport properties of hard sphere gases

a. HEAT CONDUCTION Let a gas be introduced between a pair of parallel plates of area A separated by a distance d and differing in temperature by ΔT. When a steady state has been reached, a linear temperature gradient will be established in the gas if the thermal conductivity coefficient, κ, is constant over the temperature range ΔT. The temperature of the gas at a distance z from the cool plate (temperature T_0) will then be $T = T_0 + \Delta T(z/d)$. Let the component of the free path of a gas molecule in the z-direction be λ_z. As gas molecules move, by random thermal motion, along the positive or negative z-directions, they will collide with other molecules in layers a distance λ_z apart, and differing in temperature by $\Delta T(\lambda_z/d)$. Assume for the moment that on each collision a given molecule (call it "molecule A") acquires, on the average, the mean thermal energy of the layer in which it collides. (This assumption turns out to be only roughly correct; its modification will be discussed in Section 5-6.) Suppose that, after colliding in some layer, a, of the gas, the molecule A moves away from the cool plate, colliding the next time in a new layer, b, at a distance λ_z farther from the cool plate. It is apparent that, just before molecule A suffers a collision in layer b, it will have a mean thermal energy less than that of other molecules in layer b. Therefore, when molecule A collides with a molecule in layer b, acquiring the mean thermal energy of layer b, it must absorb some energy from the layer. Similarly, if molecule A happens to move toward the cool plate, it will carry with it thermal energy in excess of that of the average molecules in the layers nearer to the cool plate. Thus molecular motion toward the hot plate results in energy absorption from the hotter layers, whereas molecular motion toward the cool plate results in the addition of energy to the cooler layers. That is, the component of random molecular motion parallel to the z-direction results in the transport of thermal energy across each layer of gas from the hot plate toward the cool plate. According to the kinetic theory of gases, it is this energy transport that is responsible for the phenomenon of heat conduction.

The magnitude of this transport of thermal energy through a given layer is found as follows. Assume for simplicity that all molecules move with the same mean free path, λ [given by Eq. (5-36)], and with the same velocity, $\bar{u}$, the mean thermal velocity. (These assumptions are, of course, not strictly valid and introduce a small error in the final result. The rigorous treatment will be discussed in Section 5-6.) Assume further that the direction of motion is random. Consider those molecules which move in the solid angle $d\Omega$ about the direction having polar angles relative to the z-direction in the range θ, $\theta + d\theta$, ϕ, $\phi + d\phi$. The number of molecules moving in this direction is [cf. Eq. (5-9)]

$$dn = \frac{n}{4\pi} d\Omega = \frac{n}{4\pi} \sin\theta \, d\theta \, d\phi \tag{5-71}$$

The component of their velocity in the z-direction is

$$u_z = \bar{u} \cos\theta \tag{5-72}$$

and the z-component of the distance between successive collisions is

$$\lambda_z = \lambda \cos\theta \tag{5-73}$$

The number of molecules moving in this direction and crossing a layer of area A in unit time is $A u_z \, dn$. Each molecule carries with it from one collision to the next an increment of thermal energy $c_v \Delta T(\lambda_z/d)$, where c_v is the heat capacity per molecule, discussed in Chapter 3. Thus the rate of transport of thermal energy due

MOLECULAR COLLISIONS

to the molecules moving in the solid angle $d\Omega$ is

$$dq = (A u_z \, dn) \left(\frac{c_v \Delta T \lambda_z}{d} \right)$$

$$= \frac{nA}{4\pi} \bar{u} \cos \theta \sin \theta \, d\theta \, d\phi \, c_v \lambda \cos \theta \, \frac{\Delta T}{d}$$

$$= \frac{n\bar{u}\lambda c_v}{4\pi} \frac{A \Delta T}{d} \cos^2 \theta \sin \theta \, d\theta \, d\phi \tag{5-74}$$

Integrating over-all values of θ and ϕ we obtain for the total rate of heat transport across a layer of area A

$$q = \frac{n\bar{u}\lambda c_v}{4\pi} \frac{A \Delta T}{d} \int_0^{2\pi} d\phi \int_0^{\pi} \cos^2 \theta \sin \theta \, d\theta$$

$$= \frac{1}{3} n\bar{u}\lambda c_v \frac{A \Delta T}{d} \tag{5-75}$$

Comparison of this result with Eq. (5-46) shows that the coefficient of heat conduction is given by

$$\kappa = \tfrac{1}{3} n \bar{u} c_v \lambda \tag{5-76}$$

Replacing the mean free path by Eq. (5-36), the mean velocity by Eq. (4-56) and the heat capacity per molecule by the molar heat capacity, $C_V = N_0 c_v$, where N_0 is Avogadro's number, we find

$$\kappa = \frac{2}{3} \left(\frac{RT}{\pi^3 M} \right)^{1/2} \frac{C_V}{\sigma^2} \tag{5-77}$$

which leads to the surprising prediction that the heat conductivity of a gas should, at a given temperature, not vary with the pressure —all of the quantities in Eq. (5-77) being independent of the pressure. This is indeed observed to be the case provided that the pressure is neither too large nor too small. The physical reason for the pressure independence of κ is the compensating effect of pressure on the mean free path and on the number of molecules per unit volume that carry the thermal energy. At high pressures there are more carriers but each carrier transports less heat because it moves a shorter distance between collisions.

It has been mentioned that several rather dubious assumptions have been made in deriving Eqs. (5-76) and (5-77) so that this result should be regarded as merely a first approximation to the correct relationship between the properties of hard sphere molecules and the heat conductivity. The modifications of the assumptions that are required will be discussed later in this chapter, but for the moment it will be sufficient to state that the rigorous theory for hard spheres leads to the expression

$$\kappa = \frac{25\pi}{64} \bar{u} n c_v \lambda \tag{5-78}$$

$$\kappa = \frac{25}{32} (RT/\pi M)^{1/2} \frac{C_V}{\sigma^2} \tag{5-79}$$

Thus the rigorous theory replaces the factor $\frac{1}{3}$ in Eq. (5-76) by a factor $25\pi/64$.

b. VISCOSITY Let a gas be placed between two parallel horizontal plates of area A set at a distance d apart, the lower plate being stationary and the upper plate sliding with a velocity U in the horizontal direction as in Fig. 5-4a. We can assume that a velocity gradient is established in the gas, so that a layer of gas at a distance z from the stationary plate moves in the same direction as the sliding plate with a velocity $U(z/d)$. Assume that on the average a molecule suffering a collision in a given layer will pick up the horizontal velocity component of mass motion, $U(z/d)$, of that layer. (This assumption is only roughly true and will be discussed further in Section 5-6.) On passing from one layer to the next, each molecule (mass m) transports momentum in the amount $mU\lambda_z/d$ for the same reason that it transports heat energy $c_v \Delta T \lambda_z/d$ in the process of heat conduction.

Let us focus our attention on a particular layer, α, of the gas. Because of their random thermal motions, molecules will move into this layer both from above and from below. Those molecules which move from below into α (i.e., molecules moving away from the stationary plate, in the positive z-direction) will, on the average, have smaller horizontal components of velocity than does the layer α because they underwent their last collisions in the more slowly moving layers. On the other hand, those molecules which move into α from above have larger horizontal velocity components than layer α because they had their last collisions in the more rapidly moving layers. Thus molecules which move into layer α from below will, on collision in the layer, be accelerated in the horizontal direction, whereas those molecules moving into α from above will be decelerated. We may say that molecules moving from below accept momentum from α and molecules moving from above donate momentum to α. The net result is that, because of the random thermal motions, momentum is carried downward through the gas across the layer α.

The same arguments show that molecules which strike the lower, stationary plate have their velocity component in the U-direction decreased (they are moving, of course, in the negative z-direction) and those molecules which strike the upper, moving plate (and are moving in the positive z-direction) have this velocity component increased. Thus forces must act between the gas and the upper and lower plates; these forces are parallel to U but act in opposite directions on the two plates. The magnitude of the force is equal to the rate of momentum transport across any layer of gas.[1]

This rate of momentum transport may be calculated in the same fashion as the rate of heat conduction was calculated above, the quantity $c_v \Delta T$ in Eqs. (5-74) and (5-75) being replaced by mU. The resulting expression for the shear force acting to maintain the velocity difference U between the two plates is

$$f = \frac{1}{3} \frac{n\bar{u}\lambda m A U}{d} \qquad (5\text{-}80)$$

Comparison with Eq. (5-56) shows that the viscosity of the gas is

$$\eta = \frac{1}{3} n\bar{u}\lambda m \qquad (5\text{-}81)$$

$$\eta = \frac{2}{3\pi^{3/2}} \frac{(MRT)^{1/2}}{N_0 \sigma^2} \qquad (5\text{-}82)$$

where N_0 is Avogadro's number.

[1] The physical origin of the shear forces in viscous flow may perhaps be visualized more effectively by means of the following example. Suppose that two trains on adjacent, parallel tracks move with slightly different velocities, ΔU, and that the passengers of the trains toss objects weighing m pounds apiece at each other at the rate of N objects per second. Then the more rapidly moving train will tend to be decelerated and the slower train will be accelerated by a force $Nm \Delta U$. In order to maintain their relative velocity at ΔU, the locomotive of the faster train will have to pull with a larger force and the locomotive of the slower train will have to pull with a diminished force. The two trains would appear to exert a viscous "drag" on one another.

Here, too, our assumptions are somewhat in error and it is found that a rigorous derivation, to be discussed in Section 5-6, yields the result

$$\eta = \frac{5\pi}{32} n\bar{u}\lambda m \tag{5-83}$$

$$\eta = \frac{5}{16\sqrt{\pi}} \frac{(MRT)^{1/2}}{N_0 \sigma^2} \tag{5-84}$$

Evidently the viscosity of a gas should, like the heat conductivity, be independent of its pressure at a given temperature—a most unexpected result which was predicted theoretically by Maxwell, and subsequently confirmed experimentally for pressures in the range of about 10^{-3} to 100 atm. Furthermore, Maxwell noted that since $\bar{u}$ increases with temperature, and since m and σ^2 should be independent of the temperature, the theory [in the form of Eq. (5-84)] predicts that the viscosity of a gas will increase with increasing temperature. This result is also surprising, because our everyday experience indicates that liquids and solids tend to flow more easily (i.e., become less viscous) when warmed. Once again experimental measurements confirmed Maxwell's prediction; indeed, the increase in the viscosity of gases with temperature is actually more pronounced than would be expected from Eq. (5-84), as will be discussed below.

c. DIFFUSION Let n_1 and n_2 be the number densities of the two components of a binary mixture of gases. Let the composition of the mixture vary in the z-direction. If the pressure is everywhere constant and if the mixture is ideal, then

$$\begin{aligned} n_1 + n_2 &= \text{const} \\ \frac{dn_1}{dz} &= \frac{-dn_2}{dz} \end{aligned} \tag{5-85}$$

The number of molecules moving in the range of directions θ, $\theta + d\theta$, ϕ, $\phi + d\phi$ and crossing the area A of a plane at $z = 0$ in the positive z-direction in unit time is

$$\left(\frac{A}{4\pi}\right) n_1^* \bar{u}_1 \cos \theta \sin \theta \, d\theta \, d\phi$$

where n_1^* is the concentration of molecules of type 1 at the layer in which they have had their last collision. If n_1 is the concentration at $z = 0$, then since the average molecules of type 1 under consideration suffered their last collisions at $z = \lambda_1 \cos \theta$, where λ_1 is the mean free path,

$$n_1^* = n_1 - \lambda_1 \cos \theta \frac{dn_1}{dz} \tag{5-86}$$

The total number of molecules of type 1 passing through $z = 0$ in the positive z-direction is therefore

$$n_1^+ = \frac{A\bar{u}_1}{4\pi} \int_0^{\pi/2} \left[n_1 - \lambda_1 \cos \theta \frac{dn_1}{dz} \right] \cos \theta \sin \theta \, d\theta \int_0^{2\pi} d\phi$$

$$= \frac{1}{4} A n_1 \bar{u}_1 - \frac{1}{6} A \lambda_1 \bar{u}_1 \frac{dn_1}{dz} \tag{5-87}$$

Similarly the number of molecules of type 1 passing through the $z = 0$ plane in the negative z-direction in unit time is

$$n_1^- = \frac{1}{4} A n_1 \bar{u}_1 + \frac{1}{6} A \lambda_1 \bar{u}_1 \frac{dn_1}{dz} \tag{5-88}$$

so that the net number of molecules passing through $z = 0$ in unit time is

$$j_1 = n_1^+ - n_1^- = -\frac{1}{3} \lambda_1 \bar{u}_1 A \frac{dn_1}{dz} \tag{5-89}$$

Comparing Eq. (5-89) with Eq. (5-59) and expressing both j_1 and c_1 in the latter equation in terms of molecules per unit time and molecules per unit volume, respectively, we see that

$$D_{12} = \frac{1}{3} \lambda_1 \bar{u}_1 \tag{5-90}$$

Introducing Eq. (5-38) for λ we obtain

$$D_{12} = \frac{1}{3\pi} \frac{\bar{u}_1}{n_1 \sigma_{11}^2 \sqrt{2} + n_2 \sigma_{12}^2 [1 + (m_1/m_2)]^{1/2}} \tag{5-91}$$

where m_1 and m_2 are the molecular masses of 1 and 2 and σ_{11} and σ_{12} are the collision diameters of molecule 1 with molecules 1 and 2. respectively. If 1 and 2 are of similar molecular size and mass and if $n = n_1 + n_2$, the total number of molecules per unit volume, we obtain

$$D_{12} = \frac{\bar{u}}{3\sqrt{2}\,\pi\sigma^2 n}$$

$$= \frac{2}{3\pi^{3/2}} \left(\frac{RT}{M}\right)^{1/2} \frac{1}{\sigma^2 n} \tag{5-92}$$

The rigorous theory produces the result, for this case

$$D_{12} = \frac{3\sqrt{2}\,\pi}{64} \bar{u}\lambda$$

$$= \frac{3}{8\sqrt{\pi}} \left(\frac{RT}{M}\right)^{1/2} \frac{1}{\sigma^2 n} \tag{5-93}$$

In the rigorous theory, the factor $\frac{1}{3}$ in Eq. (5-90) is replaced by $3\sqrt{2}\,\pi/64 = 0.208$.

Evidently the diffusion coefficients of gases should vary inversely with the pressure if the temperature is held constant—that is, in proportion to the mean free path. This is observed.

d. THERMAL DIFFUSION The molecular theory of thermal diffusion cannot be satisfactorily discussed in terms of the collision model that has been described here. The results of the application of the more rigorous theory of Chapman and Enskog to the phenomenon of thermal diffusion will be described in Section 5-6d, below.

e. TRANSPORT PROPERTIES AT VERY LOW PRESSURES The theories outlined in the previous sections all assume that the mean free path is considerably smaller than the dimensions of the vessel containing the gas, so that a molecule collides much more frequently with other molecules than with the walls of the container. We have seen that at very low pressures (generally of the order of 10 μ or below) this assumption is no longer true. The theory of transport under these conditions was developed by Knudsen. Consider the rate of transfer of heat from a wall at temperature T_1 through a gas whose mean free path is so great that its temperature, T_0, can be assumed to be uniform. If there are n molecules per unit volume, then the number of molecules whose velocity is in the range u, $u + du$ impinging on unit area of the wall in unit time is $\frac{1}{4}unP(u)\,du$ [cf. eq. (5-32)] and the rate at which translational energy is brought to the wall by these molecules is the above quantity multiplied by $\frac{1}{2}mu^2$. Using the Maxwell-Boltzmann distribution for $P(u)$, we find on integrating over all values of u that the rate at which the gas brings its own translational energy to the wall is

$$Q = \frac{1}{8} mn \int_0^\infty u^3 P(u)\, du = \frac{1}{8} mn\overline{u^3} \qquad (5\text{-}94)$$

Since $\overline{u^3} = (4/\sqrt{\pi})(2kT_0/m)^{3/2}$ and $\bar{u} = (8kT_0/\pi m)^{1/2}$ and the pressure of the gas is $p = nkT_0$, we obtain

$$Q = \frac{1}{2} p\bar{u} \qquad (5\text{-}95)$$

Let us assume that on each collision the molecules, on the average, leave the wall with translational energy $\frac{3}{2}kT_1$, corresponding to the temperature of the wall. Then the net rate of transfer of energy from the wall to the gas will be

$$q = \frac{T_1 - T_0}{T_0} Q = p\bar{u}\,\frac{T_1 - T_0}{2T_0} \qquad (5\text{-}96)$$

Experimental observations show that the actual rate of transfer for monatomic gases is less than this, so it must be recognized that only a fraction, a, of the translational energy may be transferred on a collision with the wall. (The constant a is known as the *accommodation coefficient*.) Furthermore, if the gas contains diatomic or polyatomic molecules, we must recognize that there may be a transfer of internal energy (rotation and vibration) as well as translational energy from the surface to the gas molecules on each collision. Therefore an additional factor, f, the ratio of the total heat capacity of the gas to the translational contribution to the heat capacity, must also be introduced in dealing with gases that contain more than one atom. Thus, we must write, in general,

$$q = afp\bar{u}\,\frac{T_1 - T_0}{2T_0} \qquad (5\text{-}97)$$

where a is a number between 0 and 1 and f is a number between 1 and $2A - \frac{5}{3}$ for linear molecules and between 1 and $2(A - 1)$ for nonlinear molecules, A being the number of atoms in the molecule.

Studies of the transport of heat from metals by various simple gases reveal that the accommodation coefficient, a, usually lies in the range 0.2 to 1.0.

Let us now suppose that the gas is present as a layer between two parallel plates at temperatures T_1 and T_2. We shall also suppose that the mean free path of the gas molecules is much greater than the separation of the plates. Then we can consider that there are two streams of gas moving between the plates, one with a temperature T_1' which moves from plate 1 to plate 2, the other with a temperature T_2' which moves from plate 2 to plate 1. If a is the accommodation coefficient of the plates, then from the flow of gas at plate 1 we have

$$T_2' - T_1' = a(T_2' - T_1) \tag{5-98}$$

and from the flow of gas at plate 2

$$T_1' - T_2' = a(T_1' - T_2) \tag{5-99}$$

Combining Eqs. (5-98) and (5-99), we obtain

$$T_2' - T_1' = \frac{a}{2-a}(T_2 - T_1) \tag{5-100}$$

The arguments leading to Eq. (5-96) show that, if the molecules of the gas are polyatomic, then the heat flow away from unit area of plate 2 is

$$q = \frac{1}{2} f p \bar{u} \frac{T_2' - T_1'}{T_1'}$$

or from Eq. (5-100)

$$q = \frac{1}{2} \frac{a}{2-a} f p \bar{u} \frac{T_2 - T_1}{T_1'} \tag{5-101}$$

Thus if the mean free path is greater than the separation of the plates the heat flow is proportional to the gas pressure. Furthermore, the heat flow is independent of the distance between the plates.

For the flow of gas through a tube at low pressures, Knudsen assumed that the velocities of gas molecules in the tube would be given by the Maxwell-Boltzmann distribution on which is superimposed a drift velocity, u_0, down the tube, from the end of the

MOLECULAR COLLISIONS

tube at the higher pressure p_1 to the end that is at the lower pressure p_2. The total number of molecules striking unit area of the walls of the tube in unit time is $\frac{1}{4}n\bar{u}$. If each molecule carries with it on the average a drift momentum mu_0 which it gives up to the walls on each collision, then the rate of transfer of momentum to the walls of the tube is $(2\pi RL)(\frac{1}{4}n\bar{u}mu_0)$, where R is the radius of the tube and L is its length so that $2\pi RL$ is the area of the tube wall. This rate of momentum transfer must be equated to the difference in force exerted on the two ends of the tube by the difference in gas pressures, $\pi R^2(p_1 - p_2)$. Thus

$$\pi R^2(p_1 - p_2) = \frac{\pi}{2} nm\bar{u}u_0 RL \tag{5-102}$$

$$u_0 = \frac{2R(p_1 - p_2)}{Lnm\bar{u}} \tag{5-103}$$

If r is the number of moles of gas flowing through the tube in unit time, then

$$r = \frac{\pi R^2 u_0 n}{N_0}$$

$$= \frac{2\pi R^3 (p_1 - p_2)}{M\bar{u}L} \tag{5-104}$$

where $M = N_0 m$, the molecular weight of the gas. Note that the rate of Knudsen flow of a gas through a tube is proportional to the cube of the radius whereas for viscous flow the rate is proportional to the fourth power of the radius. The flow rate under a constant pressure difference decreases as the temperature is raised because increasing the temperature increases $\bar{u}$ in the denominator of Eq. (5-103).

EXERCISE According to Poiseuille's law, the volume rate of flow of a gas flowing through a tube of radius R and length L is

$$\frac{dV}{dt} = \frac{R^4}{8L\eta} (p_1 - p_2)$$

where η is the viscosity of the gas. Since the volume of a gas depends on the gas pressure, this equation has meaning only if p_1 and p_2 are not very different. (a) Show that for large differences in p_1 and p_2 the flow rate through the tube in moles per unit time is

$$r_P = \frac{dn}{dt} = \frac{R^4}{16\pi L\eta N_0 kT}(p_1^2 - p_2^2)$$

where N_0 is Avogadro's number and k is Boltzmann's constant. (b) If we write r_K for the flow rate in Knudsen flow, as given in Eq. (5-104), show that the ratio of the flow rate that the gas has under Knudsen conditions to the flow rate that it would have if it flowed according to Poiseuille's law is

$$\frac{r_K}{r_P} = \frac{32\pi^2}{3}\frac{\lambda_1}{R}\frac{p_1}{p_1 + p_2}$$

where λ_1 is the mean free path of the gas at pressure p_1. Thus at very low pressures, where $\lambda \gg R$, the Knudsen flow rate is much greater than one would expect if the gas were flowing as a viscous fluid with the normal value of the viscosity coefficient.

EXERCISE Show that if the flow rate of a gas through a tube is expressed in terms of the volume V_1 of gas flowing into the tube (and measured at the entrance pressure, p_1), then the Poiseuille flow rate is

$$\frac{dV_1}{dt} = \frac{R^4}{16\pi L\eta}\left(p_1 - \frac{p_2^2}{p_1}\right)$$

whereas the Knudsen flow rate is

$$\frac{dV_1}{dt} = \frac{2\pi R^3 kT}{mL\bar{u}}\left(1 - \frac{p_2}{p_1}\right)$$

where m is the molecular mass and k is Boltzmann's constant. Note that for Knudsen flow under these circumstances the maximum volume flow rate of the gas through the tube (i.e., the rate when the exit pressure, p_2, is zero) is independent of the entrance pressure.

5-5 Experimental tests of the hard sphere theories of transport properties

a. ESTIMATION OF COLLISION DIAMETERS FROM TRANSPORT PROPERTIES Table 5-5 gives the observed values of the transport properties η, κ, and D for some typical gases, along with the values of σ^2, the square of the collision diameter, calculated by means of both the approximate theory [Eqs. (5-77), (5-82), and (5-92)] and the more rigorous theory to be discussed below [Eqs. (5-79), (5-84),

TABLE 5-5 TESTS OF HARD SPHERE THEORIES OF TRANSPORT PROPERTIES OF GASES[a]

Gas	$10^5 \eta$ (g/cm sec)	$10^5 \kappa$ (cal/cm sec deg)	D (cm^2/sec)	Square of collision diameter, σ^2 (in Å^2)									Sutherland constant °K
				Approx. theory			Rigorous theory						
				from η	from κ	from D	from η	from κ	from D	$M\kappa/C_v\eta$	$\rho D/\eta$	$1 + \frac{9}{4}\frac{R}{C_v}$	
He	19.20	34.06	—	2.21	1.32	—	4.61	4.85	—	2.37	—	2.50	80
Ne	29.67	11.10	0.452	3.35	1.80	3.30	6.98	6.60	5.83	2.53	1.375	2.50	56
A	20.99	3.94	0.156	6.47	3.63	6.82	13.5	13.3	12.05	2.51	1.325	2.50	142
Kr	23.27	2.08	0.081[b]	8.31	4.74	9.08	17.6	17.4	16.05	2.50	1.30	2.50	188
Xe	21.07	1.23	0.048[b]	11.5	6.39	12.2	24.0	23.5	21.6	2.57	1.33	2.50	252
H$_2$	8.53	39.65	1.285	3.51	2.64	3.68	7.35	9.68	6.50	1.895	1.49	1.91	84
D$_2$	11.8	30.31	—	3.60	2.45	—	7.50	9.00	—	2.07	—	1.91	—
N$_2$	16.63	5.50	0.185	6.75	5.04	6.85	14.1	18.5	12.1	1.89	1.39	1.92	104
O$_2$	19.18	5.84	0.187	6.26	4.49	6.35	13.1	16.5	11.2	1.975	1.39	1.91	125
Air	17.08	5.79	0.181	6.65	4.75	6.92	13.9	17.45	12.25	1.99	1.365	1.92	112
CO$_2$	13.66	3.49	0.0974	10.30	9.05	10.42	21.5	33.2	18.45	1.655	1.40	1.66	254
H$_2$O	8.61	4.29	—	10.47	9.78	—	21.9	35.6	—	1.51	—	1.76	650
CH$_4$	10.30	7.21	0.206	8.23	6.28	8.15	17.2	23.0	14.4	1.87	1.43	1.75	164
C$_2$H$_6$	8.51	4.36	—	13.65	13.4	—	28.5	49.0	—	1.45	—	1.43	252
n-C$_4$H$_{10}$	6.86	3.22	—	23.4	26.5	—	48.9	97.0	—	1.26	—	1.21	358
n-C$_6$H$_{14}$	5.86	2.96	—	33.2	35.0	—	69.4	128.5	—	1.34	—	1.14	436
HCl	13.13	—	0.1063[b]	9.77	—	10.4	20.4	—	18.45	—	1.32	—	362

[a] Except where otherwise indicated, data are for gases at 0°C and 1 atm. Data from Hirschfelder, Curtiss, and Bird, *Molecular Theory of Gases and Liquids*, John Wiley and Sons, New York, 1954, and Partington, *An Advanced Treatise on Physical Chemistry*, Vol. 1, John Wiley and Sons, New York, 1949.

[b] Values extrapolated from data at slightly higher temperatures (20–30°C).

and (5-93)]. The following conclusions may be reached from an inspection of Table 5-5:

(1) Since molecular diameters are known to be of the order of a few Ångstrom units (1 Å = 10^{-8} cm) for the simple gases in Table 5-5, and since the values of σ^2 in this table are all of the order of 10^{-16} cm^2, it is evident that the theory described in Section 5-4 must be at least a good first approximation to reality.

(2) There is excellent agreement between the values of σ^2 derived from the viscosity, η, and from the self-diffusion coefficient, D, using the approximate theory of Section 5-4.

(3) The value of σ^2 derived from the approximate theory of the heat conductivity, κ, is usually considerably smaller than the values obtained from the approximate theories of the viscosity and diffusion coefficients.

(4) The values of σ^2 obtained from the rigorous theory for the three properties, η, κ, and D, for the monatomic gases are in excellent agreement. The discrepancy noted in the previous paragraph is therefore caused in some part by the assumptions introduced into the approximate theory.

(5) The values of σ^2 derived from the rigorous hard sphere theory of the heat conductivity of polyatomic gases are larger than those obtained from the rigorous, hard sphere theories of viscosity and diffusion. This is especially noticeable in the series of hydrocarbons, CH_4, C_2H_6, n-C_4H_{10} and n-C_6H_{14}, where values of σ^2 obtained from the heat conductivity are approximately twice as large as the values obtained from the viscosity. The reason for this failure of the rigorous theory will be discussed in Section 5-6.

b. RELATIONSHIPS BETWEEN η, κ, AND D According to the approximate theories of the heat conductivity and viscosity, Eqs. (5-76) and (5-81), we should expect that the ratio of these two properties would be given by

$$\frac{\kappa}{\eta} = \frac{C_v}{M} \qquad (5\text{-}105)$$

so that the approximate theory predicts

$$\frac{M\kappa}{C_v\eta} = 1 \qquad (5\text{-}106)$$

It can be seen from Eqs. (5-78) and (5-83) that the rigorous theory for hard spheres leads to the prediction

$$\frac{M\kappa}{C_V\eta} = \frac{5}{2} \tag{5-107}$$

Similarly Eqs. (5-90) and (5-81) lead to the prediction, according to the approximate theory,

$$\frac{D}{\eta} = \frac{N_0}{Mn} \tag{5-108}$$

Since n is the number of molecules per unit volume, and M/N_0 is the molecular mass, the quantity Mn/N_0 is the density, ρ, of the gas, so we obtain the prediction, according to the approximate theory

$$\frac{\rho D}{\eta} = 1 \tag{5-109}$$

From the rigorous theory for hard spheres [Eqs. (5-93) and (5-83)] we find

$$\frac{\rho D}{\eta} = \frac{6}{5} \tag{5-110}$$

Observed values of $M\kappa/C_V\eta$ and $\rho D/\eta$ are given in Table 5-5. For the monatomic gases the former quantity is close to 2.50, the value predicted by the rigorous theory, and the latter quantity is in the range 1.3 to 1.4, which is well above the value predicted by the approximate theory (1.0) and somewhat above the value predicted by the rigorous theory (1.20). For polyatomic gases both $M\kappa/C_v\eta$ and $\rho D/\eta$ deviate somewhat from the value predicted by the rigorous theory, showing that the model of these gases as perfectly rigid spheres leaves something to be desired. This will be discussed further in Section 5-6.

c. VARIATION OF THE VISCOSITY AND HEAT CONDUCTIVITY AT HIGH PRESSURE The theories that we have been considering predict that the viscosity and heat conductivity of a gas should be independent of the pressure at constant temperature. We have seen that at very low pressures, where the mean free path becomes comparable with the physical dimensions of the apparatus used to study the transport properties, this is no longer the case. We should also expect the theories to fail at high pressures, where the mean free path becomes comparable with molecular dimensions.

Figure 5-6 shows the variation of the viscosity of nitrogen with pressure at 50°C. Below 50 atm the viscosity is constant to within better than 5%, but at higher pressures there is an appreciable increase; at 700 atm the viscosity is double the low pressure

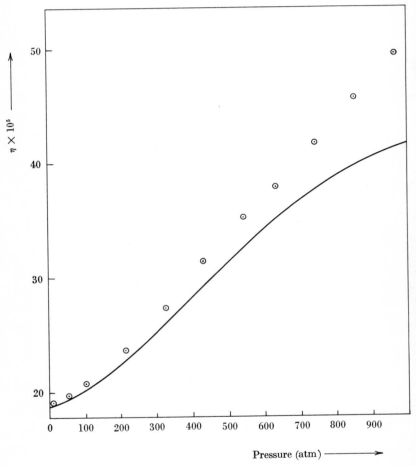

FIG. 5-6 *Variation of the viscosity of nitrogen with pressure at 50°C. Solid curve is calculated from Eq. (5-113). Circles give experimental values found by Michels and Gibson [Proc. Roy. Soc. (London),* **A134**, *288 (1931)].*

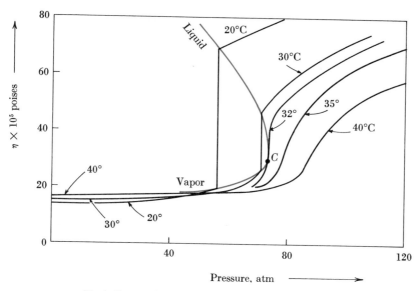

FIG. 5-7 *Variation of the viscosity of carbon dioxide with pressure and temperature near the critical point. The critical point is indicated by the point C. The critical temperature is 32°C. Colored line indicates the liquid-vapor equilibrium locus.* [From P. Phillips, Proc. Roy. Soc. (London), **A87**, 48 (1912).]

value. Figure 5-7 shows the variation of the viscosity of carbon dioxide in the vicinity of the critical point. In general the viscosity increases considerably when a gas achieves densities comparable to that of the liquid state, and the pressure variation is not as great at high temperatures as it is at low temperatures.

The thermal conductivity of gases also increases at high pressures, the behavior for some simple gases being shown in Fig. 5-8. Evidently the pressure variation is small up to a few tens of atmospheres.

d. VARIATION OF TRANSPORT PROPERTIES WITH TEMPERATURE According to the hard sphere theories, the heat conductivity and the viscosity of gases should be proportional to the square root of the temperature. Since the diffusion coefficient is inversely proportional to the density as well as directly proportional to the square root of the temperature [cf. Eqs. (5-92) and (5-93)] and since the density at constant pressure is inversely proportional to the temperature we should expect that

$$D_p \propto T^{3/2}$$

D_p is the diffusion coefficient measured at a constant pressure.

Figure 5-9 shows the observed temperature variation of $\eta/\sqrt{MT}$ and $D_p \sqrt{M}/T^{3/2}$ at a pressure of 1 atm. It is evident that the hard sphere model predicts a more gradual variation in the viscosity and diffusion coefficient than is actually observed, since both $\eta/\sqrt{T}$ and $D_p/T^{3/2}$ increase somewhat with increasing temperature whereas the hard sphere theories predict that both quantities should be independent of temperature.

It is found empirically that over moderate ranges of temperature the viscosity-temperature relation is given quite well by the expression

$$\eta = AT^n$$

where A and n are constants, n having values in the range 0.7 to

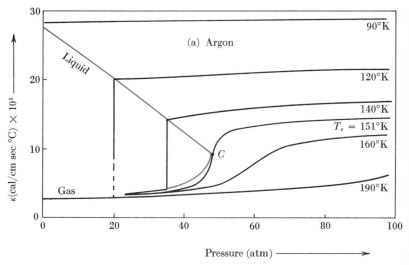

FIG. 5-8 (a) *Variation of the thermal conductivity with pressure for some simple gases. Argon in the vicinity of the critical point C.* [*Curves are approximate and taken from data of A. Uhlir, J. Chem. Phys.,* **20,** *463 (1952).*]

1.0 for the common gases. Similar relations are also obeyed by the thermal conductivity and the diffusion constant at constant density. We shall find that useful information can be obtained about the forces acting between molecules from these small deviations from the predictions of the hard sphere theories.

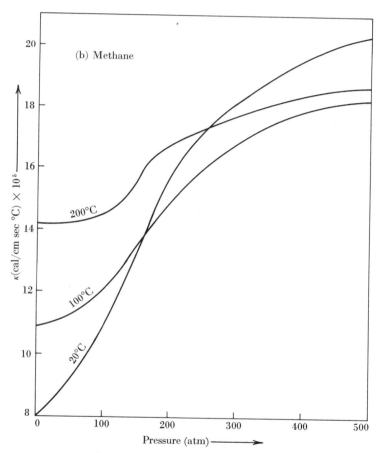

FIG. 5-8 (b) Variation of the thermal conductivity with pressure for some simple gases. Methane above room temperature. [From E. A. Stoljarow, W. W. Ipatjew, and W. P. Teodorowitsch, J. Phys. Chem. (USSR), 24, 166 (1950).]

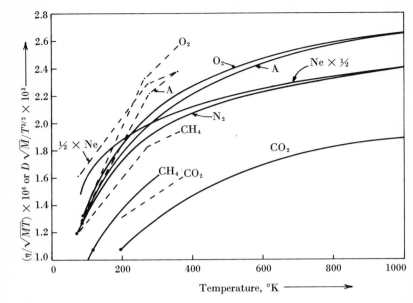

FIG. 5-9 *Temperature variation of the transport properties of simple gases. Solid lines show the variation of $\eta/(MT)^{1/2}$ with temperature. Dashed lines show the variation of $D_p(M/T^3)^{1/2}$ with temperature. Both $\eta/(MT)^{1/2}$ and $D_p(M/T^3)^{1/2}$ are inversely proportional to the collision cross section. Filled circles show the boiling points of the gases. (Data from Tables 8.4-2 and 8.4-13 of J. O. Hirschfelder, C. F. Curtiss, and R. B. Bird, Molecular Theory of Gases and Liquids, Wiley, New York, 1954.)*

5-6 *Refinement of the molecular theory of transport properties*

a. EFFECT OF FINITE MOLECULAR SIZE ON TRANSPORT PROPERTIES AT HIGH DENSITIES Consider a straight row of N identical, stationary, rigid spheres of diameter σ, equally spaced with their centers at a distance λ from each other. Let the sphere at the left end of the row be given a velocity u toward the right. On collision with its neighbor, the velocity u will be transferred to the neighbor, which in turn will hand on the velocity u to its neighbor and so on down the row. Eventually, the sphere at the right-hand end of the row will acquire the velocity u. If the diameters of the spheres were negligible in comparison with the distance between the spheres, the time required for the impulse to be transmitted from one end of the row to the other would be $N\lambda/u$. Since, however, the spheres are assumed to be perfectly rigid, the impulse travels across each sphere with infinite velocity, so that if the finite size of the spheres is taken into account, the time required for the impulse to travel down the row is $N(\lambda - \sigma)/u$. If the number of spheres in unit length of the row is $n = 1/\lambda$, we see that the time required for the impulse to move through unit length of the row is

$$\tau = \frac{1}{u}(1 - n\sigma) \tag{5-111}$$

so that the effective velocity for the transport of the impulse through the row is

$$u_{\text{eff}} = \frac{u}{(1 - n\sigma)} \tag{5-112}$$

If gas molecules can be regarded as rigid spheres, a similar effect is to be expected in gases at high densities. As the density of molecules increases, the rate of transport of impulses and disturbances such as those in viscosity and heat conduction through the gas can be expected to increase above the value that would obtain if this effect of finite molecular size were neglected. Thus it can be predicted that the viscosity and heat conductivity of a gas should increase as the gas density increases. We have seen that this is indeed the case (Figs. 5-5, 5-6 and 5-7).

A theory of this effect for a three-dimensional gas consisting of rigid spheres of diameter σ was worked out by Enskog, whose theory leads to the relation

$$\eta = \eta_0 \left(1 + 0.175 \frac{b\rho}{M} + 0.865 \frac{b^2\rho^2}{M^2}\right) \tag{5-113}$$

where ρ is the gas density, M is the molecular weight of the gas, and b is the van der Waals constant defined in Chapter 2, $b = (2\pi/3)N_0\sigma^3$, N_0 being Avogadro's number. Equation (5-113) is found to agree well with the experimental observations. The observed pressure dependence of the viscosity of nitrogen ($b = 57.0$ cm^3/mole from $P - V$ data, $M = 28$) is compared with that calculated from Eq. (5-113) in Fig. 5-6 using $\eta_0 = 18.8 \times 10^5$.

The agreement, while not perfect, is satisfactory and indicates that Enskog's rigid sphere model for nitrogen at high pressures may have at least an element of truth behind it. It is believed that a number of additional, more subtle effects may also operate, however.[2]

b. EFFECT OF ATTRACTIVE FORCES AND MOLECULAR "SOFTNESS" ON THE TEMPERATURE DEPENDENCE OF THE TRANSPORT PROPERTIES OF GASES   The hard sphere model cannot be entirely correct because we know from the equation of state of real gases, liquids, and solids that molecules attract one another at large distances and that there is no precisely defined interatomic distance within which it is impossible for two molecular centers to approach each other. Thus the concept of a collision diameter as applied to real gases can be only approximately correct.

It is pertinent at this point to investigate the meaning of the concept of a collision between two molecules. Let two molecules of masses m_1 and m_2 approach each other in the manner indicated in Fig. 5-10a, where it is assumed that the center of gravity of the two molecules is stationary and that the motion is confined to the plane of the paper. Successive positions of the two molecules during the encounter are indicated by the serial numbers 1, 2, . . . , 12. Since the center of gravity is at rest, the paths of the molecules before the encounter must be parallel, and at each instant the centers of the two particles and the center of gravity must lie along a common straight line. The ratio of the distances from the center of gravity to the two particles must be constant and equal to the inverse of the ratio of the corresponding masses. Thus if the trajectory of one particle is known, that of the other can easily be constructed. In Fig. 5-10a it is assumed that the particles at first attract each other on their approach, so that initially their paths are drawn together. The molecules ultimately approach each other sufficiently closely so that strong repulsive forces are set up. The paths then begin to diverge. Finally, when the molecules have moved out of the region in which they influence each other appreciably, they will move apart along parallel paths that make an angle χ with the directions of the initial paths. The

---

[2] Kim and Ross [*J. Chem. Phys.*, 42, 263 (1965)] give reasons for doubting that the "collisional transfer" effect considered by Enskog is the only important factor in determining the density dependence of the viscosity. They show that dimers arising from intermolecular attractions probably make an important contribution to the effect.

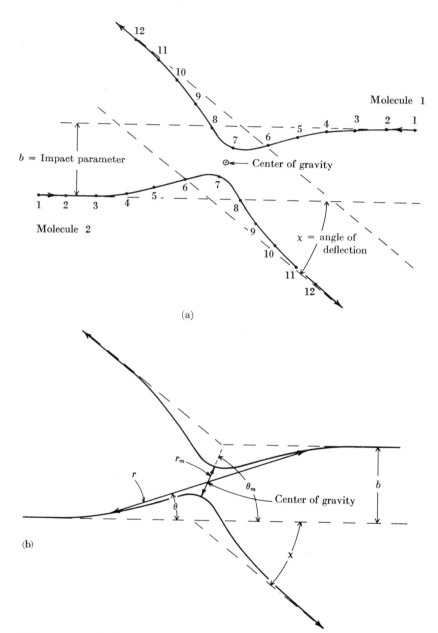

FIG. 5-10 *Paths of a pair of attracting molecules during a collision.*

angle χ, which is called the *angle of deflection*, is a measure of the violence of the collision.

The distance b between the extensions of the initial paths of the two molecules is known as the *impact parameter*. (The impact parameter is thus the minimum distance at which the molecules would pass each other if no forces existed between them so that the encounter had no effect on their motion.) It is obvious that the angle of deflection, χ, will depend upon the impact parameter b—large values of b giving small values of χ, whereas, if $b = 0$, the directions of motion are reversed so that $\chi = 180°$. Furthermore, χ may, for real molecules, depend on the speed with which the two molecules involved in an encounter approach one another; if for a given value of the impact parameter the molecules approach at high speed, they will usually be deflected to a lesser extent than if they approach at low speed.

It is not difficult to obtain a relationship which shows how the angle of deflection depends on the impact parameter, the relative velocity and the intermolecular forces. Let x_1, y_1, z_1 and x_2, y_2, z_2 be the Cartesian coordinates of two atoms, the origin being at the center of gravity. Then the kinetic energy of the two atoms is

$$T = \frac{1}{2}[m_1(\dot{x}_1^2 + \dot{y}_1^2 + \dot{z}_1^2) + m_2(\dot{x}_2^2 + \dot{y}_2^2 + \dot{z}_2^2)] \tag{5-114}$$

where m_1 and m_2 are the respective masses and $\dot{x}_1$ is the velocity component of particle 1 in the x-direction, and similarly for $\dot{x}_2, \dot{y}_1$, etc. If no external forces act on the atoms, their centers must move in a plane. It is therefore possible to rotate the coordinate axes in such a way that the motion is expressed by but two coordinates for each atom. We therefore lose nothing by restricting the motion to the xy-plane, so that $\dot{z}_1 = \dot{z}_2 = 0$. Furthermore, let the distance between the particles be r and let θ be the angle between the line joining the particles (which passes through the origin) and the direction of motion before the encounter—which will be taken to be the direction of the x-axis (see Fig. 5-10b). The position of the center of gravity having been fixed at the origin of the coordinate system, it is necessary that

$$m_2 x_1 = -m_1 x_2 \tag{5-115}$$
$$m_2 y_1 = -m_1 y_2$$

MOLECULAR COLLISIONS

Furthermore

$$x_1 - x_2 = r \cos \theta$$
$$y_1 - y_2 = r \sin \theta \tag{5-116}$$

From these four relationships we obtain the equations

$$x_1 = \frac{m_2}{m_1 + m_2} r \cos \theta$$

$$y_1 = \frac{m_2}{m_1 + m_2} r \sin \theta$$

$$x_2 = -\frac{m_1}{m_1 + m_2} r \cos \theta \tag{5-117}$$

$$y_2 = -\frac{m_1}{m_1 + m_2} r \sin \theta$$

Thus the four variables, x_1, y_1, x_2, y_2, may be expressed in terms of the two variables, r and θ. Taking time derivatives we find

$$\dot{x}_1 = \frac{m_2}{m_1 + m_2} (\dot{r} \cos \theta - r\dot{\theta} \sin \theta)$$

$$\dot{y}_1 = \frac{m_2}{m_1 + m_2} (\dot{r} \sin \theta + r\dot{\theta} \cos \theta)$$

$$\dot{x}_2 = -\frac{m_1}{m_1 + m_2} (\dot{r} \cos \theta - r\dot{\theta} \sin \theta) \tag{5-118}$$

$$\dot{y}_2 = -\frac{m_1}{m_1 + m_2} (\dot{r} \sin \theta + r\dot{\theta} \cos \theta)$$

which, on substitution in Eq. (5-114) gives

$$T = \frac{1}{2} \mu (\dot{r}^2 + r^2 \dot{\theta}^2) \tag{5-119}$$

where μ is the reduced mass,

$$\mu = \frac{m_1 m_2}{m_1 + m_2} \tag{5-120}$$

If the potential energy of the system depends only on the separation of the two molecules, and is written as $V(r)$, and if we disregard contributions to the energy by internal motions (vibrations and rotations) of the molecules, then the total energy is

$$E = \frac{1}{2}\mu(\dot{r}^2 + r^2\dot{\theta}^2) + V(r) \tag{5-121}$$

The motion of the two molecules is equivalent to the motion of a single molecule whose mass is equal to the reduced mass, μ, and is subjected to a potential field $V(r)$. If $V(r) = 0$ when the molecules are far apart, and if g is the initial relative speed, then since $\dot{\theta} = 0$ and $\dot{r} = g$ when the molecules are far apart, we have for the total energy before the collision

$$E = \frac{1}{2}\mu g^2 \tag{5-122}$$

It is assumed that no external forces are acting on the system and that the collisions are elastic, so that the energy must be constant and we must have at all times

$$\frac{1}{2}\mu g^2 = \frac{1}{2}\mu(\dot{r}^2 + r^2\dot{\theta}^2) + V(r) \tag{5-123}$$

Furthermore, the angular momentum about the center of gravity cannot change because no external torques are assumed to act on the system. Thus the system must move in such a way that

$$\mu b g = \mu r^2 \dot{\theta} \tag{5-124}$$

where $\mu b g$ is the angular momentum about the center of gravity before the molecules begin to interact. Substitution of Eq. (5-124) into Eq. (5-123) gives

$$\dot{r}^2 = g^2\left(1 - \frac{b^2}{r^2}\right) + \frac{2}{\mu}V(r) \tag{5-125}$$

Let r_m be the distance of closest approach in the collision and let θ_m be the angle θ at the point of closest approach (Fig. 5-10b). The motion as the molecules move apart is exactly the reverse of their motion on approaching each other, so that θ_m must be related to the angle of deflection, χ, by

$$\chi = \pi - 2\theta_m \tag{5-126}$$

MOLECULAR COLLISIONS

The angle θ_m can be found for a given pair of values of g and b in the following way. Note that

$$\frac{dr}{d\theta} = \frac{dr/dt}{d\theta/dt} = \frac{\dot{r}}{\dot{\theta}} \tag{5-127}$$

Substituting from Eqs. (5-124) and (5-125) we obtain

$$\frac{\dot{r}}{\dot{\theta}} = \pm \frac{r^2}{b}\left(1 - \frac{b^2}{r^2} - \frac{2V(r)}{\mu g^2}\right)^{1/2} \tag{5-128}$$

where the negative sign is to be used before the molecules have reached the point of closest approach and the positive sign is to be used after this time ($\dot{\theta}$ changes in the same sense at all times during the collision, but $\dot{r}$ is negative before the collision and positive after the collision.) Thus up to the point of closest approach

$$d\theta = -\frac{b}{r^2}\frac{dr}{\sqrt{1 - (b^2/r^2) - (2V/\mu g^2)}} \tag{5-129}$$

On integration we find

$$\theta_m = \int_0^{\theta_m} d\theta = -b \int_\infty^{r_m} \frac{1}{r^2} \frac{dr}{\sqrt{1 - (b^2/r^2) - (2V/\mu g^2)}} \tag{5-130}$$

so that from Eq. (5-126) we obtain for the dependence of the angle of deflection on the impact parameter b and on the initial relative velocity g

$$\chi(b, g) = \pi + 2b \int_\infty^{r_m} \frac{dr}{r^2 \sqrt{1 - (b^2/r^2) - (2V/\mu g^2)}} \tag{5-131}$$

This equation permits us to calculate the angle of deflection resulting from the collision of a pair of molecules of known interaction potential $V(r)$ for any pair of values of the impact parameter, b, and initial relative velocity, g.

Equation (5-131) is readily evaluated for hard, nonattracting spheres. In this case we may write $V(r) = 0$ if $r > \sigma$, and $V(r) = \infty$ if $r < \sigma$. If the impact parameter, b, is less than the collision diameter, σ, a collision will take place in which the distance of closest approach is $r_m = \sigma$. Since $V(r) = 0$, the angle of deflection does not depend on the relative velocity g, and we may write

$$\chi(b) = \pi + 2b \int_\infty^\sigma \frac{dr}{r^2 \sqrt{1 - (b^2/r^2)}} \qquad (b \leq \sigma) \qquad (5\text{-}132)$$

The integral may be evaluated by replacing the variable r by the variable $z = b/r$, giving

$$\chi(b) = \pi - 2 \int_0^{b/\sigma} \frac{dz}{\sqrt{1 - z^2}}$$

$$= \pi - 2 \sin^{-1}(b/\sigma) = 2 \cos^{-1}(b/\sigma) \qquad (b \leq \sigma) \qquad (5\text{-}133)$$

On the other hand, if the impact parameter is greater than the collision diameter, the distance of closest approach will be equal to the impact parameter itself and we have

$$\chi(b) = \pi + 2b \int_\infty^b \frac{dr}{r^2 \sqrt{1 - (b^2/r^2)}}$$

$$= \pi - 2 \int_0^1 \frac{dz}{\sqrt{1 - z^2}}$$

$$= \pi - 2 \sin^{-1} 1 = 0 \qquad (b \geq \sigma) \qquad (5\text{-}134)$$

confirming that under these circumstances no deflection will take place.

It is also possible to evaluate the integral in Eq. (5-131) for particles which interact according to an inverse square law of force, such as a pair of charges e_1 and e_2, for which the intermolecular potential is

$$V(r) = \frac{e_1 e_2}{r} \qquad (5\text{-}135)$$

This potential gives

$$\chi = \cos^{-1} \frac{\epsilon^2 - 1}{\epsilon^2 + 1} \qquad (5\text{-}136)$$

where

$$\epsilon = \frac{bg^2 \mu}{e_1 e_2} \qquad (5\text{-}137)$$

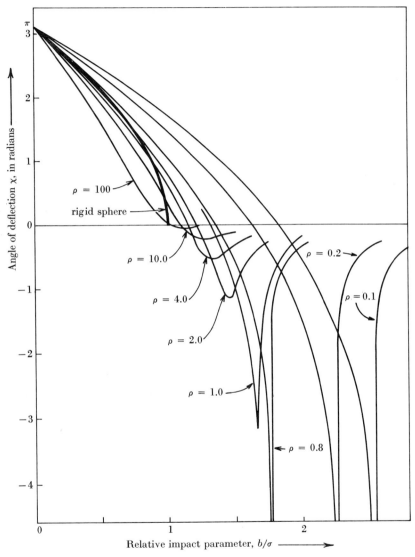

FIG. 5-11 *Dependence of the angle of deflection on the impact parameter and relative initial velocity for rigid spheres and for spherical molecules obeying the Lennard-Jones potential. (Numerical values from J. O. Hirschfelder, C. F. Curtiss, and R. B. Bird, Molecular Theory of Gases and Liquids, Wiley, New York, 1954, pp. 1132 ff.)*

The dependence of χ on b and g has been calculated by Hirschfelder, Bird, and Spotz [3] for the Lennard-Jones 6-12 potential (which has been described in Section 2-3a)

$$V(r) = 4\epsilon \left[\left(\frac{\sigma}{r}\right)^{12} - \left(\frac{\sigma}{r}\right)^{6} \right] \tag{5-138}$$

where ϵ and σ are empirical constants. As shown in Chapter 2, this potential has the general shape expected for a pair of real molecules. The results of the calculations of Hirschfelder, Curtiss, and Bird are shown in Fig. 5-11. Curves of χ vs. b/σ are shown for various values of ρ, the ratio of the kinetic energy of relative motion, to ϵ, the depth of the trough in the intermolecular potential,

$$\rho = \frac{1}{2} \frac{\mu g^2}{\epsilon} \tag{5-139}$$

Also shown in Fig. 5-11 is the relationship between b and χ for a pair of rigid, nonattracting spheres. It is evident that if a pair of molecules approach one another with a very high velocity (large values of ρ), they tend to be deflected as if they were rigid spheres with collision diameters approximately equal to σ. For small relative velocities (small values of ρ), however, the variation of χ with b is more complex, violent deflections being obtained for b values that are considerably larger than the distance parameter σ. If ρ is less than about 1.0, and if the impact parameter falls within a particular (but very narrow) range of values, it is found that the particles can make one or more complete rotations about their center of gravity. Sketches of some approximate molecular trajectories at low and high values of ρ are shown in Fig. 5-12.

These results show that the mean free path (i.e., the average distance travelled by a molecule between successive violent deflections of its path) used in the simple theories of transport outlined in previous sections of this chapter cannot be independent of the temperature, as was assumed in these theories; for real molecules, the effective collision diameter must depend considerably on the relative velocity of the colliding pair. Inspection of Fig. 5-11

[3] *J. Chem. Phys.*, **16**, 968 (1948) and Table I-R, pp. 1132 ff. of J. O. Hirschfelder, C. F. Curtiss, and R. B. Bird, *Molecular Theory of Gases and Liquids*, Wiley, New York, 1954.

reveals that when the temperature is raised from a value at which $\rho = 1$ for an average molecule (this being slightly below the temperature, T_b, at which the substance boils) to a value at which $\rho = 10$ for an average molecule (i.e., to a temperature of approximately $10\ T_b$), the range of impact parameters b over which the angle of deflection is greater than, say $10°$ (about 0.2 rad), decreases from 2.0 σ for $\rho = 1.0$ to about 1.3 σ when $\rho = 10$. Since the mean free path varies inversely with the square of the collision diameter, this means that between the boiling temperature and ten times the boiling temperature the mean free path might increase by a factor of something like $(2.0\sigma/1.3\sigma)^2 = 2.4$, the density of the gas being held constant. Therefore the quantities $\eta/\sqrt{T}$ and $\rho D/T^{3/2}$, which according to the hard sphere model ought to be independent of the temperature, should in fact increase by roughly a factor of something like 2 or 3 between T_b and $10 T_b$. A variation of roughly this magnitude is indeed observed for real gases. In Fig. 5-9 the quantity $\eta/\sqrt{T}$ is plotted against the temperature over a range of temperatures that extends from approximately the boiling point to ten times the boiling point, the boiling points being indicated on the curves by large dots. The values of $\eta/\sqrt{T}$ at ten times this temperature are seen to be roughly double or triple the values at the boiling point.

Thus we are led to believe that the existence of attractive forces between molecules and the absence of a sharply defined distance at which repulsive forces set in may be responsible for a large part of the temperature variation of the transport properties of gases, variations that should not exist according to the hard sphere model developed earlier in this chapter. We are also led to the conclusion that information about intermolecular forces should be obtainable from a study of the temperature dependence of the transport properties of gases.

In 1893 Sutherland developed an approximate theory of the transport properties of gases which is fairly successful in reproducing their temperature dependence. He assumed that a gas molecule is a rigid sphere of diameter σ, but that for intermolecular distances greater than σ there exists an attractive potential of the form

$$V(r) = -\epsilon \left(\frac{\sigma}{r}\right)^n \qquad (r > \sigma) \qquad (5\text{-}140)$$

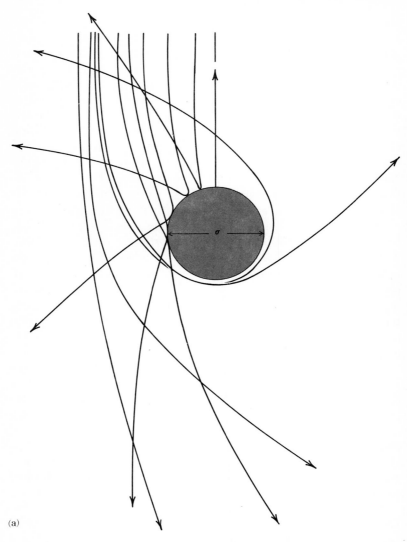

(a)

FIG. 5-12 *Approximate trajectories relative to a stationary center of gravity for collisions involving a pair of molecules interacting by a Lennard Jones 6–12 potential. One of the molecules approaches from the top of the figure with various impact parameters. The center of gravity of the pair is at*

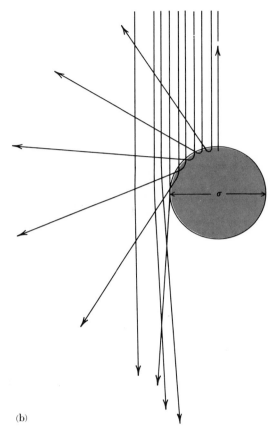

(b)

FIG. 5-12 *(Continued)*
the center of the colored circle, which has a diameter σ, where σ is the length parameter of the Lennard-Jones potential. The other molecule involved in the collision approaches from the bottom of the figure; its trajectories are not shown, but they may be found by inverting the trajectories of the first molecule through the center of gravity. (a) Trajectories for low velocity collisions. The two molecules approach with kinetic energy $\frac{1}{2}\mu g^2$ equal to 0.1ϵ, where ϵ is the energy parameter of the Lennard-Jones potential and g is the relative velocity of the molecules when they are far apart. (b) Trajectories for high-velocity collisions, $\frac{1}{2}\mu g^2 = 50\epsilon$.

where ϵ and n are positive constants. He was able to show that at high temperatures this leads to the viscosity

$$\eta = \frac{\eta_0}{1 + (S/T)} \quad (5\text{-}141)$$

where η_0 is the viscosity one would have if the gas consisted of nonattracting hard spheres of radius σ. The constant S, known as Sutherland's constant, is given by

$$S = \frac{i(n)\epsilon}{nk} \quad (5\text{-}142)$$

k being Boltzmann's constant and $i(n)$ being a number whose value depends on n, but is not far from 0.2 when n lies in the range 2 to 8. Empirical values of the Sutherland constant for several gases are given in Table 5-5. These are useful in estimating the variation of the viscosity of gases over a limited range of temperatures. As expected, S tends to be largest for gases with the highest boiling points [largest attractive potential, hence the largest value of ϵ in Eqs. (5-140) and (5-142)].

Chapman showed that if the molecules of a gas interact with a potential of the form

$$V(r) = \frac{K}{r^n} \quad (5\text{-}143)$$

then the temperature variation of the viscosity and heat conductivity should be given by a relationship of the form

$$\eta = \eta_0 \left(\frac{T}{T_0}\right)^\nu \quad (5\text{-}144)$$

$$\kappa = \kappa_0 \left(\frac{T}{T_0}\right)^\nu \quad (5\text{-}145)$$

where

$$\nu = \frac{1}{2} + \frac{2}{n} \quad (5\text{-}146)$$

and where η_0 and κ_0 are the viscosity and heat conductivity at temperature T_0. Equation (5-144) gives an excellent description of the observed temperature dependence of the viscosity over restricted ranges of temperatures; values of ν observed for simple gases such as air, carbon dioxide, and the rare gases around room temperature are in the range 0.65 to 1.0 corresponding to n values in the range 5 to 12.

C. REFINEMENTS IN THE THEORY OF HEAT CONDUCTION; PROBLEM OF THE RATE OF TRANSFER OF INTERNAL ENERGY BY COLLISION It has been noted in Section 5-5 that the collision diameters of diatomic and polyatomic gases calculated from the hard sphere theory are out of line with the values obtained from the viscosity and the diffusion constant. In the hard sphere theory of heat conduction as outlined in Section 5-4, it was assumed that the efficiency of transfer of translational thermal energy is the same as the efficiency of transfer of rotational and vibrational thermal energy. We have seen in Table 5-5 that for the rare gases, which possess only translational energy, the quantity $M\kappa/C_V\eta$ has the value $\tfrac{5}{2}$ predicted by the exact hard sphere theory rather than the value unity predicted by the approximate theory. Eucken reasoned that one should write

$$\kappa = \epsilon_{\text{trans}} \frac{\eta C_{V\,\text{trans}}}{M} + \epsilon_{\text{int}} \frac{\eta C_{V\,\text{int}}}{M} \tag{5-147}$$

where $C_{V\,\text{trans}}$ and $C_{V\,\text{int}}$ are the contributions of translational motion and internal motions to the molar heat capacity, and ϵ_{trans} and ϵ_{int} are appropriate numerical factors. The exact theory demonstrates that ϵ_{trans} must be given the value $\tfrac{5}{2}$. Eucken suggested that the assumptions of the approximate theory which led to a value $\epsilon_{\text{trans}} = 1$ should apply to the contributions of the internal motions, so that he set $\epsilon_{\text{int}} = 1$. Since $C_{V\,\text{trans}} = \tfrac{3}{2}R$ and $C_{V\,\text{int}} = C_V - \tfrac{3}{2}R$, this leads to the relation

$$\kappa = \frac{\eta}{M}\left[\frac{5}{2}\cdot\frac{3}{2}R + C_V - \frac{3}{2}R\right] = \frac{\eta}{M}\left[C_V + \frac{9}{4}R\right] \tag{5-148}$$

or

$$\frac{\kappa M}{\eta C_V} = 1 + \frac{9}{4}\frac{R}{C_V} \tag{5-149}$$

Values of the quantity $1 + (9R/4C_V)$ are given in Table 5-5 and are seen to be in fairly good agreement with the observed values of

$\kappa M/\eta C_V$, especially for the diatomic molecules. (Of course, they must agree well for the monatomic gases, for which $C_V = \frac{3}{2}R$ and $1 + (9R/4C_V) = 2.50$.) The agreement is not, however, impressive for most of the polyatomic gases.

The reason for this failure for the polyatomic gases is believed to lie in the poor efficiency with which vibrational energy is transferred on collisions between molecules. There is much evidence that hundreds and even thousands of collisions are required to transfer a quantum of vibrational energy from one molecule to another, whereas only a very few collisions are required to change the translational or rotational energy of a molecule to the extent required by the heat conduction process. The detailed corrections to the theory of heat conductivity have not yet been worked out, however.

d. RIGOROUS THEORIES OF TRANSPORT PROPERTIES In the simple derivations of equations for the transport properties presented in Section 5-4, several assumptions have been made which, while not grossly wrong, cannot be precisely correct. For instance, it was assumed that each molecule, on collision with another molecule at a given point in the gas, acquires on the average the local mass velocity and temperature of that point. Furthermore, it was assumed that all molecules have the same free path and the same random velocity, and the derivations were based on the picture of a molecule as a rigid sphere of definite collision diameter, which as we have seen is only a rough first approximation to the truth. The rigorous theories which make it possible to deal correctly with these shortcomings are uncommonly difficult, and no effort will be made to derive them here. The basis for these theories was laid by Maxwell in 1866, when he considered the problem of the modification of the distribution function of a gas which has not reached a completely random state. For the next fifty years, many theoreticians wrestled with the problem of relating the transport properties of gases to the forces acting between molecules, but the practical solution of the problem was only achieved in the years 1911–1917 by Chapman and Enskog, working independently. It was the theory developed by these workers that established the correct value of the numerical coefficients for hard sphere gases given in Eqs. (5-78), (5-83) and (5-93).

The results of the Chapman-Enskog theory can be conveniently presented by defining the weighted cross sections (note that they have the dimensions of an area)

$$Q_1(g) = 2\pi \int_0^\infty (1 - \cos \chi) b \, db \tag{5-150}$$

$$Q_2(g) = 2\pi \int_0^\infty (1 - \cos^2 \chi) b \, db \tag{5-151}$$

where χ is the deflection angle discussed earlier, b is the impact parameter, and g is the relative velocity of a pair of molecules before they have come close to one another. We may recall that the deflection angle, χ, depends on b and g and that χ approaches zero as b becomes large. Thus the factors $(1 - \cos \chi)$ and $(1 - \cos^2 \chi)$ in the above integrals go to zero when b is larger than the distance at which a pair of molecules begin to have an influence on one another. These factors assume values of the order of unity for values of b which lead to violent collisions. The quantities $Q_1(g)$ and $Q_2(g)$ can therefore be considered as two types of average cross-sectional areas of the colliding molecules. If we define further the quantities

$$\Omega_{mn}(T) = \sqrt{1/4\pi} \int_0^\infty e^{-\mu g^2/2kT} \left(\frac{\mu g^2}{2kT}\right)^{2n+3} Q_m(g) \, dg \tag{5-152}$$

then it is found that the transport properties of pure gases are given by

$$\eta = \frac{5kT}{8\Omega_{22}} \tag{5-153}$$

$$\kappa = \frac{25kT}{16\Omega_{22}} \tag{5-154}$$

$$D = \frac{3k^2 T^2}{8mP\Omega_{11}} \tag{5-155}$$

It will be recalled that the deflection angle χ depends on the intermolecular potential, so it is evident that the magnitude of the quantities Ω_{mn} depend directly on this potential. For hard spheres of collision diameter σ it is found that

$$Q_1 = \pi \sigma^2 \tag{5-156}$$

$$Q_2 = \tfrac{2}{3}\pi \sigma^2 \tag{5-157}$$

$$\Omega_{11} = \sqrt{2\pi\mu/kT} \, \pi\sigma^2 \tag{5-158}$$

$$\Omega_{22} = \sqrt{2\pi\mu/kT} \, 2\pi\sigma^2 \tag{5-159}$$

When substituted in Eqs. (5-153), (5-154) and (5-155) these expressions lead to the results given in Eqs. (5-79), (5-84) and (5-93).

Hirschfelder *et al.* have evaluated these functions for molecules whose interactions are given by the Lennard-Jones potential. By fitting the observed viscosity over a wide range of temperatures to their calculated viscosity vs. temperature curves, they have been able to derive values of the empirical constants ϵ and σ in the Lennard-Jones potential. Some typical results are shown in Table 2-1. The agreement with the parameters obtained from second virial coefficient data is excellent.

Another result of the Chapman-Enskog theory is the evaluation of the thermal diffusion coefficient. The expression for mixtures of different chemical species is quite complex, but for a pair of isotopic species of molecular weights M_1 and M_2 it is found that the thermal diffusion factor [defined in Eq. (5-70)] is

$$\alpha = \frac{15(2A + 5)(6C - 5)}{2A(16A - 12B + 55)} \frac{M_1 - M_2}{M_1 + M_2} \tag{5-160}$$

where the quantities A, B, and C are defined by

$$A = \frac{1}{2} \frac{\Omega_{22}}{\Omega_{11}} \tag{5-161}$$

$$B = \frac{5\Omega_{12} - \Omega_{13}}{3\Omega_{11}} \tag{5-162}$$

$$C = \frac{\Omega_{12}}{3\Omega_{11}} \tag{5-163}$$

It is found that for hard spheres,

$$\Omega_{12} = 3\pi\sigma^2 \left(\frac{2\pi\mu}{kT}\right)^{1/2} \tag{5-164}$$

$$\Omega_{13} = 12\pi\sigma^2 \left(\frac{2\pi\mu}{kT}\right)^{1/2} \tag{5-165}$$

which, with the values of Ω_{11} and Ω_{22} given in Eqs. (5-158) and (5-159), give $A = B = C$, so that for the hard sphere isotopic species we find

$$(\alpha)_{\text{hard sphere}} = \frac{105}{118} \frac{M_1 - M_2}{M_1 + M_2} \tag{5-166}$$

The observed values of α for isotopic mixtures are somewhat smaller than this, but they tend to approach the hard sphere value at high temperatures. Thus for $\text{Ne}^{20} - \text{Ne}^{22}$ containing 90% of the lighter isotope, the ratio $(\alpha)_{\text{obs}}/(\alpha)_{\text{hard sphere}}$ is 0.382 at 129°K and 0.816 at 712°K.[4]

Problems

1. The heat capacity, C_V, of nitrogen is 5.0 cal/deg mole and its viscosity at room temperature is 1.7×10^{-4} poises. How much heat will be conducted in one second across a 1 mm space between two parallel plates 10 cm × 10 cm in size if the plates differ in temperature by 5°C and if the space between the plates is filled with nitrogen at 1 atm?

2. The viscosities of halogen vapors are as follows:

Halogen	Temperature (°C)	Viscosity (poises × 10^5)
Chlorine	100	16.9
Bromine	286	15.1
Iodine	124	18.4

Find the collision diameters.

3. Sulfur hexafluoride (SF_6) and adamantane ($C_{10}H_{16}$) are both very nearly spherical molecules which do not differ greatly in molecular weight but which do differ somewhat in size. From a study of bond lengths and van der Waals radii one would expect the collision diameter of SF_6 to be about 6.06 Å. and that of adamantane should be about 7.80 Å. Estimate the viscosities of these two substances in the vapor state at 400°C.

4. Two gases are composed of molecules that have the same volumes and the same masses, but the molecules of one gas are nearly spherical whereas those of the other are elongated (e.g., cyclohexane and hexene-3, both having the formula C_6H_{12}). Which gas would you expect to have the higher viscosity? Why?

5. The molecules CH_4 and CD_4 have almost the same shape and size. At a given temperature, which substance will have the greater viscosity and which will effuse the more rapidly through a tiny hole? Why?

6. The viscosity of neon at 25°C is 31.3×10^{-5} poises. How long

[4] L. G. Stier, *Phys. Rev.*, **62**, 548 (1942)

does it take for an average neon molecule to collide with one mole of other neon molecules at 25°C and 1 atm?

7. The Tait mean free path, λ_T, is defined as the average distance traveled by a molecule before its first collision, beginning at some given random instant in time rather than at the point of its last collision. Show that

$$\lambda_T = \int_0^\infty \lambda_u P(u)\, du = \frac{1}{\pi n \sigma^2} \int_0^\infty \frac{4x^2 e^{-x^2}}{\psi(x)}\, dx$$

where $\psi(x)$ is the function defined in Eq. (5-42). [The integral in this equation is found to have the numerical value 0.677, so that $\lambda_T = 0.677/\pi n\sigma^2$, which is about 4% less than the mean free path between successive collisions given by Eq. (5-36).]

8. The mean free path in a certain gas is 1.10 mm. A given molecule happens to have moved exactly 0.80 mm since its last collision without having encountered another molecule. How much farther can one expect the molecule to travel before it has its next collision?

9. A one-liter flask initially filled with air is attached to a large vacuum pump through a tube one meter long and one centimeter in diameter. Assume that the pump is able to maintain zero pressure at the low pressure end of the tube. The viscosity of air is 17×10^{-5} poises. (a) If the flow through the tube obeys Poiseuille's law, how long will it take for the pressure in the flask to fall from 1 atm to 0.1 atm? (b) How long will it take for the pressure to fall from 10^{-3} to 10^{-4} atm under the same conditions, assuming Poiseuille flow? (c) How long will it take for the pressure to fall from 10^{-5} to 10^{-6} atm if the gas moves through the tube by Knudsen flow? (d) How long will it take for the pressure to fall from 10^{-8} to 10^{-9} atm?

10. A two-liter container is divided into two equal volumes, A and B, by means of a thin diaphram containing a hole 0.1 mm in diameter. Volume A is evacuated and B is filled with hydrogen at a pressure of 1 mm Hg. The entire system is maintained at 25°C. How long will it take for the pressure in B to drop from 1 mm Hg to 0.75 mm Hg? [*Hint:* Note Eq. (5.32).]

11. In the system described in Problem 10, nitrogen at a pressure of 1 mm Hg is introduced into A and hydrogen at a pressure of 1 mm Hg is introduced into B. Describe the subsequent pressure changes in both chambers.

12. A fluid flows between two stationary, parallel plates. Derive a relationship between the volume rate of flow of the fluid, the viscosity of the fluid, the dimensions of the plates (which are taken to be rectangular in shape), their separation and the pressure difference between the points of entrance and exit of the fluid in the space between the plates.

13. A long tube contains air at 1 atm and 0°C. A layer of hydrogen gas 1 cm thick is introduced into the center of the tube. How long will it take for the hydrogen pressure to build up by diffusion to 0.001 atm at a point one meter down the tube? The diffusion constant of hydrogen through air is 0.68 cm²/sec at 0°C and 1 atm.

14. A tube is filled with hydrogen gas containing 5.21 moles percent deuterium. One end of the tube is heated to 125°C and the other end is maintained at 25°C. What is the isotopic concentration at the hot end of the tube? The thermal diffusion factor, α, is 0.17 for hydrogen-deuterium mixtures.

15. Energy from the sun reaches the earth at the rate of 1.92 cal/min cm². A spherical satellite 10 cm in diameter is painted to give a "black" surface (emissivity = 1) which absorbs all of the energy coming from the sun. (a) What is the steady state temperature of the satellite if the only means of gaining and losing thermal energy is by radiation? (b) If instead of painting the satellite the surface is polished, so that only 1% of the sun's energy is absorbed, the emissivity remaining at unity, what will the steady state temperature be?

16. The "black" satellite in the previous problem is constructed of aluminum (mean specific heat 0.2 cal/°C g between −80°C and 500°C) and weighs 100 g. When it passes into the shadow of the earth it no longer receives energy from the sun. What will its temperature be after it remains in the earth's shadow for 30 min?

17. Verify Eqs. (5-156), (5-157), (5-158), and (5-159).

18. A spaceship approaches Mars (mass 6.3×10^{26} g) moving with a velocity of 20,000 km/sec relative to Mars and with an impact parameter of 15,000 km. (a) What is the angle of deflection of the spaceship after it has undergone its encounter with Mars? (The interaction between Mars and the spaceship is determined by the gravitational potential, $V(r) = -Gm_1m_2/r$, where m_1 and m_2 are the masses of Mars and of the spaceship, r is the distance between the two objects and $G = 6.66 \times 10^{-8}$ erg − cm/g².) (b) The diameter of Mars is 6,800 km. What would the impact parameter of the above-described spaceship have to be if it were to suffer a grazing collision with Mars' surface? [*Hint:* When $r = r_m$, $\dot{r} = 0$ in Eq. (5-125).]

Supplementary references

T. G. Cowling, *Molecules in Motion*, Harper Torchbooks, New York, 1960. A readable elementary account of the kinetic theory and its applications.

R. D. Present, *Kinetic Theory of Gases* (see References, Chapter

2). Contains a quite detailed, lucid and up-to-date account of the transport properties of gases, from both hard sphere and Chapman-Enskog points of view.

E. A. Guggenheim, *Elements of the Kinetic Theory of Gases*, Pergamon, New York, 1960. Especially good for its brief but clear account of the Chapman-Enskog theory.

J. O. Hirschfelder et al., *Molecular Theory* (see References, Chapter 2). Chapters 7–11 contain an exhaustive and definitive account of the Chapman-Enskog theory, with extensive application to experimental observations.

S. Chapman and T. G. Cowling, *The Mathematical Theory of Non-Uniform Gases*, Cambridge, 1939 (available in a paperback edition). This is the classical account of the Chapman-Enskog theory.

The older texts by Loeb, Jeans, and Kennard (see References, Chapter 2) are of interest and value. They emphasize the hard sphere theories.

The properties of gases at very low pressures are discussed in the following:

M. Knudsen, *The Kinetic Theory of Gases*, Methuen, London, 1934. Informal and lucid.

S. Dushman, *Scientific Foundations of Vacuum Technique*, 2nd ed. (J. M. Laffery, ed.), Wiley, New York, 1962, Chapters 1 and 2. See also Chapter 7 of Loeb and Chapter 8 of Kennard.

Thermal diffusion in gases:

K. E. Grew and T. L. Ibbs, *Thermal Diffusion in Gases*, Cambridge, 1952. See also the German revision of this monograph, which includes more recent results, by the same authors, *Thermodiffusion in Gasen*, Veb. Deutschen Verlag der Wissenschaften, 1962.

Measurement of diffusion constants in gases:

W. Jost, *Diffusion in Solids, Liquids, Gases*, revised ed., Academic Press, New York, 1960, pp. 406–413.

Appendix 5-1

EVALUATION OF THE INTEGRALS IN EQUATION (5-22)

WE MUST evaluate the sixfold integral

$$I = \iiint\iiint_{-\infty}^{\infty} \exp\left(-\frac{m_A(u_x^2 + u_y^2 + u_z^2) + m_B(u_x'^2 + u_y'^2 + u_z'^2)}{2kT}\right) u_{\text{rel}}\, du_x\, du_y\, du_z\, du_x'\, du_y'\, du_z' \quad (5\text{-}1\text{-}1)$$

where

$$u_{\text{rel}} = [(u_x - u_x')^2 + (u_y - u_y')^2 + (u_z - u_z')^2]^{1/2} \quad (5\text{-}1\text{-}2)$$

Introduce the new variables

$$U_x = \frac{m_A u_x + m_B u_x'}{m_A + m_B}$$

$$U_y = \frac{m_A u_y + m_B u_y'}{m_A + m_B}$$

$$U_z = \frac{m_A u_z + m_B u_z'}{m_A + m_B}$$

$$u_{rx} = u_x - u'_x \qquad (5\text{-}1\text{-}3)$$
$$u_{ry} = u_y - u'_y$$
$$u_{rz} = u_z - u'_z$$
$$U^2 = U_x^2 + U_y^2 + U_z^2$$
$$u_{\text{rel}}^2 = u_{rx}^2 + u_{ry}^2 + u_{rz}^2$$
$$\mu = \frac{m_A m_B}{m_A + m_B}$$

Note that

$$m_A(u_x^2 + u_y^2 + u_z^2) + m_B(u_x'^2 + u_y'^2 + u_z'^2) = (m_A + m_B)U^2 + \mu u_{\text{rel}}^2 \quad (5\text{-}1\text{-}4)$$

Note also that U_x, U_y, and U_z are the velocity components of the center of mass of the colliding molecules. Now, it can be shown that

$$du_x\, du'_x = dU_x\, du_{rx}$$
$$du_y\, du'_y = dU_y\, du_{ry} \qquad (5\text{-}1\text{-}5)$$
$$du_z\, du'_z = dU_z\, du_{rz}$$

so that Eq. (5-1-1) may be written

$$I = \iiiint\!\!\iint e^{-[(m_A+m_B)U^2 + \mu u_{\text{rel}}^2]/2kT} u_{\text{rel}}\, dU_x\, dU_y\, dU_z\, du_{rx}\, du_{ry}\, du_{rz}$$

$$= \iiint e^{-(m_A+m_B)U^2/2kT} dU_x\, dU_y\, dU_z$$

$$\iiint e^{-\mu u_{\text{rel}}^2/2kT} u_{\text{rel}}\, du_{rx}\, du_{ry}\, du_{rz} \quad (5\text{-}1\text{-}6)$$

If we transform to polar coordinates in both integrals, writing

$$dU_x\, dU_y\, dU_z = U^2 \sin\theta\, dU\, d\theta\, d\phi$$
$$du_{rx}\, du_{ry}\, du_{rz} = u_{\text{rel}}^2 \sin\theta'\, du_{\text{rel}}\, d\theta'\, d\phi' \qquad (5\text{-}1\text{-}7)$$

and if we integrate over all directions θ, ϕ, θ', and ϕ' in both integrals, we obtain

$$I = 16\pi^2 \int_0^\infty U^2 e^{-(m_A+m_B)U^2/2kT}\, dU \int_0^\infty u_{\text{rel}}^3 e^{-\mu u_{\text{rel}}/2kT}\, du_{\text{rel}}$$

$$(5\text{-}1\text{-}8a)$$

APPENDIX 5-1

$$= 4\pi^{5/2}\left(\frac{2kT}{m_A + m_B}\right)^{3/2} \int_0^\infty u_{\text{rel}}^3 e^{-\mu u_{\text{rel}}^2/2kT} \, du_{\text{rel}} \qquad (5\text{-}1\text{-}8\text{b})$$

$$= 4\pi^{5/2}\left(\frac{2kT}{m_A + m_B}\right)^{3/2} \frac{1}{2}\left(\frac{2kT}{\mu}\right)^2 \qquad (5\text{-}1\text{-}8\text{c})$$

Substitution into Eq. (5-22) gives for the total number of collisions per unit time and per unit volume

$$Z_{AB} = \frac{1}{8} n_A n_B \pi \sigma_{AB}^2 \frac{(m_A m_B)^{3/2}}{(\pi kT)^3} \cdot 4\pi^{5/2}\left(\frac{2kT}{m_A + m_B}\right)^{3/2} \frac{1}{2}\left(\frac{2kT}{\mu}\right)^2$$
$$= n_A n_B \pi \sigma_{AB}^2 \, (8kT/\pi\mu)^{1/2} \qquad (5\text{-}1\text{-}9)$$

From Eq. (5-22) and Eq. (5-1-8b), above, we also see that the collision frequency for molecules whose relative speed is between u_{rel} and $u_{\text{rel}} + du_{\text{rel}}$ is

$$dz_{AB} = \frac{1}{8} n_A n_B \pi \sigma_{AB}^2 \frac{(m_A m_B)^{3/2}}{(\pi kT)^3} \left[\frac{2kT}{m_A + m_B}\right]^{3/2} 4\pi^{5/2} u_{\text{rel}}^3$$
$$\exp\left(\frac{-\mu u_{\text{rel}}^2}{2kT}\right) du_{\text{rel}}$$
$$= (2/\pi)^{1/2} n_A n_B \pi \sigma_{AB}^2 (\mu/kT)^{3/2} u_{\text{rel}}^3$$
$$\exp\left(\frac{-\mu u_{\text{rel}}^2}{2kT}\right) du_{\text{rel}} \qquad (5\text{-}1\text{-}10)$$

This result is useful in the theory of chemical reactions.

Index of Symbols

THIS INDEX lists the more important symbols used more than once, with page where defined or discussed.

a	Van der Waals' constant, 34
$B, B(T), B', B'(T)$	Second virial coefficients, 23, 32
b	Van der Waals' constant, 34
b	Impact parameter, 220
$C, C(T), C', C'(T)$	Third virial coefficients, 23, 32
C_V	Heat capacity (at constant volume), 93
D	Diffusion coefficient, diffusivity, 192
E	Thermal energy (macroscopic), 107
e	Base of natural logarithms ($= 2.71828\ldots$), 34, 81
$\exp(x)$	The base of natural logarithms raised to the power x ($= e^x$), 46
g	Acceleration of gravity, 18-19
g	Relative velocity of two gas molecules at large separation, 222
h	Planck's constant, 101
I	Moment of inertia, 102
k	Boltzmann's constant, 63
M	Molecular weight, 16
m	Molecular mass, 101
N	Number of molecules, 51

Index of Symbols

N_0 Avogadro's number, 63
n Number of moles, 16
P Pressure, 4-7
$P(u)$ Distribution function for the property u, 139
Q Partition function, 107
R Gas constant, 16
T Absolute temperature, 13
t Temperature, 13
u Molecular velocity, 51-54
V Volume, 4
$V(r)$ Intramolecular potential, 78, 81
Z Collision frequency for all molecules in a gas, 167
z Collision frequency for a single molecule, 167
z Compressibility factor, 22
α Thermal diffusion factor, 197
ϵ Energy (molecular), 107-108
η Viscosity coefficient, 188
κ Thermal conductivity, 184-185
λ Mean free path, 180
ν Molecular vibration frequency, 103
ρ Density, 18
σ Collision diameter, 167
θ Constant in the vibrational partition function ($= h\nu/k$), 111
χ Angle of deflection, 220

Index of Subjects

Absolute temperature scale, 13
Accommodation coefficient, 205
Angle of deflection, 220
Atmosphere, normal (pressure unit), 18
Average values, 154–155
Average velocity (*see* Velocity, mean)
Avogadro's law, 16, 55

Bar (pressure unit), 18
Barometric formula, 46–47
Barye (pressure unit), 18
Bernoulli theory, 50–60
Black body radiation (*see* Radiation, thermal)
Boltzmann, 97
Boltzmann constant, 63
Boltzmann factor, 105–107, 147
  derivation, 131–136, 163
Boyle's law, 10, 52, 60

Celsius scale (temperature), 13
Centigrade scale (temperature), 13
Chapman-Enskog theory of transport properties, 232–235

Charles' law, 12
Collision diameter, 167, 208–210
Collision frequency, 166–179
  detailed dynamics, 218–229
  with surfaces, 178–179
Composition variables, 8, 21
Compressibility factor, 22, 30, 32, 44
Corresponding states, 41
Covolume, 64
Critical constants, numerical values, 31–32
Critical point, 27
  van der Waals' theory, 37–39

Dalton's law, 20–21, 55–56
de Broglie wavelength, 142
Degeneracy, 127–128
Degree of freedom, 94
Diameter, collision (*see* Collision diameter)
Diffusion, effect of temperature, 213–216
  molecular theories, hard sphere theory, 202–204
  rigorous theory, 233

Diffusion, phenomenological theory, 191–196
  self-, 195
Diffusivity, 192
Dimerization, effect on equation of state, 72–73, 89
Dispersion forces, 75–77
Distribution function, 58, 137
  for energies, 161
  Maxwell-Boltzmann, 147–152
    effect of temperature, 159–160
  for velocities, 147–154

Effusion, 61
Energy, electronic, 125, 129
  equipartition principle, 96
  rotational, 96, 114–118
  translational, 91–94, 96, 119–121
  vibrational, 96, 109–114
  zero point, 103, 127
Energy levels, degenerate, 127–128
Equation of state, 8–9
  Beattie-Bridgeman, 36
  Berthelot, 34, 40, 45, 74
  Dieterici, 34, 40, 45, 74
  real gases, 30–44
  reduced, 41, 45–46
  van der Waals', 34, 37–40, 64–75
Equipartition of energy, 96
Error function, 158–159
Error integral, 158–159
Expansion, thermal, 12

Fahrenheit scale (temperature), 13
Fick's laws of diffusion, 192–193
Flux, 192
Forces, intermolecular, dispersion, 75–77
  Lennard-Jones potential, 77–79, 226, 234
  London, 75–77
  van der Waals', 70
Frequency, molecular vibration (*see* Vibration, molecular, frequency)

Gas constant, 16
  dimensions, 19
  units, 19–20

Gas laws, ideal, 10–21
Gauss error function, 158–159
Gauss error integral, 158–159
Gay-Lussac's law, 12, 55
Graham's law, 61–63
Gram-mole, 16, 18

Heat capacity, diatomic gases, 123
  effect of intermolecular forces, 127
  electronic, 125, 129
  experimental measurement for gases, 124, 130
  gases, classical theory, 96–97
  tables of values, 100
  monatomic gases, 93
  polyatomic gases, 124–125
  rotational, 117–118
  temperature variation, 125–127
  vibrational, 113
Heat conduction, effect of temperature, 209, 213–216, 227, 230
  Eucken theory, 231–232
  at high pressure, 213
  internal energy transfer, 231–232
  molecular theory, hard sphere theory, 197–200
  rigorous theory, 233
  phenomenological theory, 184–188
  by thermal radiation, 185–186
Hougen-Watson chart, 43

Ideal gas laws, 10–21
Impact parameter, 220

Kelvin scale (temperature), 13
Knudsen flow, 206–208

Lennard-Jones potential, 77–79
London forces, 75–77

Mass, 4
Maxwell-Boltzmann distribution function, 140
Mean free path, 180–184
  effect of intermolecular attractions, 218–226
  Tait, 236
Mean values, 154–155

INDEX OF SUBJECTS

Mean velocity, 53
Mole, 16
Mole fraction, 21
Moment of inertia, 102

Newton's law of flow, 188
Non-Newtonian flow, 188
Normal mode, 96

Oscillator, harmonic, quantum rules, 103

Partial pressure (see Pressure, partial)
Partition function, 107
  diatomic molecule, 121–122
  rotational, 114–118
  translational, 119–121
  vibrational, 110
Planck's constant, 101
Poise, 188
Poiseuille's law, 190–191, 207–208
Potential energy, intermolecular (see Forces, intermolecular)
Pound-mole, 16, 18
Pressure, 4
  critical, 27
  partial, 20–21, 55
  units, 17–19
  vapor, 26
Primary properties, 3
Properties, extensive, 44–45
  intensive, 44–45
  primary, 3

Quantum number space, 148
Quantum numbers, 104
Quantum rules, energy, 101–103
  velocity, 142

Radiation, thermal, 179–180
  heat conduction, 185–186
Rankine scale (temperature), 13
Real gases, equations of state, 22–44
  van der Waals' theory, 64–75
Reduced variables, 41
Rotator, quantum rules, 102

Self-diffusion, 195
Slip, 191
Stefan's law, 179
Sutherland's constant, 229

Tait mean free path, 236
Temperature, 7
  Boyle, 46, 89
  critical, 27
  kinetic theory of, 55
Temperature scales, 13
Thermal conductivity (see Heat conduction)
Thermal diffusion, molecular theory, 204
  phenomenological theory, 196–197
  rigorous theory, 234–235
Thermal expansion of gases, 12
Thermal radiation (see Radiation, thermal)
Torr (pressure unit), 18
Translational energy, 91–94
Transport properties, 184–197
  hard sphere theory, 192–204
  rigorous theory, 232–235
  at very low pressures, 204–208

Vapor pressure, 26
Variables, composition, 8, 21
  primary, 3
  reduced, 41
Velocity, average, 53
  mean, 53, 155–156
  median, 157
  most probable, 156
  rms, 61, 157
  root-mean-square, 61, 156–157
Velocity quantization, 142
Vibration, molecular, frequency, 103, 104, 124
Vibrational mode, 95–96
Virial, coefficients, 23–25, 30–36, 45
  statistical mechanical theory, 81–88
  temperature variation, 24, 33, 35–36
  equation of state, 23–25, 30–36

Viscosity, effect of temperature, 209, 213–216, 227, 230, 231
  at high pressure, 211–213, 216–218
  molecular theory, hard sphere theory, 200–202
    rigorous theory, 233
  phenomenological theory, 188–191
  Sutherland's equation, 227, 230

Volume, 4
  critical, 27
  excluded, 64
  units, 18

Waterston, 50, 97

Zero point energy, 103